D0152098

WEATHERING
An introduction to the scientific principles

Will Bland
School of Applied Chemistry, Kingston University

David Rolls
Formerly School of Geography, Kingston University

A member of the Hodder Headline Group
LONDON • NEW YORK • SYDNEY • AUCKLAND

First published in Great Britain in 1998 by
Arnold, a member of the Hodder Headline Group
338 Euston Road, London NW1 3BH
http://www.arnoldpublishers.com

Copublished in the United States of America by
Oxford University Press Inc.,
198 Madison Avenue,
New York, NY 10016

British Library Cataloguing in Publication Data
A catalogue entry for this book is available from the British Library

Library of Congress Cataloging-in-Publication Data
A catalog entry for this book is available from the Library of Congress

ISBN 0 340 67745 7 (hb)
ISBN 0 340 67744 9 (pb)

Production Editor: James Rabson
Production Controller: Rose James
Cover Design: Terry Griffiths

Typeset in 10/12 pt Palatino by Academic & Technical, Bristol
Printed and bound in Great Britain by MPG Books Ltd, Bodmin, Cornwall

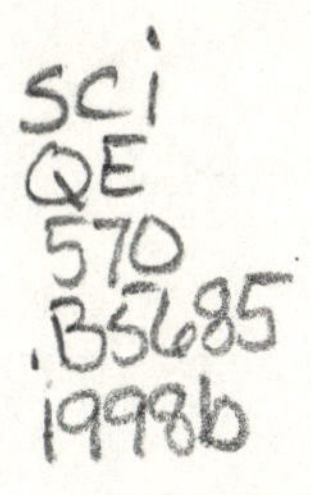

CONTENTS

PREFACE

We define weathering as 'the alteration by chemical, mechanical, and biological processes of rocks and minerals, at or near the Earth's surface, in response to environmental conditions'. This definition could have been elaborated to take into account ideas such as the view of weathering as the tendency of rocks in disequilibrium to be transformed into more stable substances. We think, however, that our definition provides a concise and useful starting position from which more detailed concepts can be developed.

In writing this book we have focused on the needs of students in the first and second years of degree courses in geography, the Earth sciences, the environmental sciences and related areas. At the same time we hope that some of the material will be useful for third-year students, and indeed anyone interested in how rocks weather at and near the Earth's land surface. We have tried to fill a perceived gap in the literature between the general account, which tends to 'gloss over' or even distort the underlying chemistry especially, and the advanced work which is often inappropriate for many undergraduates.

Weathering is a difficult topic. This is partly because of the complex nature of the subject, which involves the interaction of a very large number of factors, and partly because a proper understanding requires a knowledge of basic chemistry and physical principles. Little can be done to reduce the first problem, but an attempt is made in this book to overcome the second difficulty.

Students commonly enter Earth science and related courses with good backgrounds of learned information which has often been derived from textbooks, but only modest knowledge of scientific principles. In this book we have tried to provide an understanding of the principles behind the weathering processes. These are relatively uncontroversial where accepted chemical ideas are concerned, but explanations become increasingly uncertain when natural environments are studied. We have assumed little prior knowledge, and have tried to develop the various explanations in a logical and user-friendly manner.

A grasp of the various principles will help the student to overcome the difficulty of dealing with the complex nature of the subject. For example, the leaching of metal ions from a mineral is best understood from a knowledge of the principles of the structure of solids and the physical behaviour of ions in water.

The structure of the book develops naturally from this approach. We start by considering the nature of the Earth materials on which weathering acts. The major role of water in weathering processes leads us to consider its properties next. The weathering processes themselves, mechanical, chemical and biological, occupy the next three chapters, which are followed by a review of weathering rates and intensities. The final two chapters on the results of weathering and weathering-controlled landforms draw on the earlier work for an integrated treatment of these subjects.

In the detailed presentation we have tried to maintain our approach of arguing from first principles. We have used worked examples so that the reader can appreciate the quantitative meaning of certain processes. We have used boxes to isolate detailed material from the main text where typically only principles and their direct implications are discussed. The boxed material may be useful for study in class. Some of the topics covered are more advanced, e.g. chemical thermodynamics, rates of reaction and the relationship of Eh and pH. These are still explained from first principles, but could also be usefully studied at a later stage when an understanding of basic weathering processes has grown.

The text material is supported by various lists. Many of the more technical terms and chemical compounds we have referred to are briefly defined where they are first introduced. However, they are often used later and the reader may find a reminder of their meaning helpful. We have therefore provided two separate lists at the end of the book. One is a glossary of terms, and the other is a list of named compounds (e.g. goethite) and their chemical composition. We have also provided a list of the main units and constants which we have used. At the end of each chapter we have provided a representative list of useful readings to help the student with further study. Each reference has been lightly annotated. In general we have decided not to give references in support of the more detailed statements as this would interfere with the flow of the text.

In summary, our approach has been to focus on the scientific principles underlying weathering, in the belief that a knowledge of these provides the best way of understanding the topic, and of supporting later studies.

Will Bland
David Rolls
Kingston University

1

GENERAL PRINCIPLES OF BONDING AND STRUCTURE IN EARTH MATERIALS

Simple observation tells you that earth materials differ greatly in their physical properties. If you dig up a spadeful of soil you will see earth which has a crumbly structure mixed up with stones which are hard. In dry environments you will find rock salt crystals which are hard and brittle, or you could find sparkling micas which are hard but easily sheared. Water, also found in soil, is a liquid.

The physical differences in these materials reflect both the ways in which the sub-microscopic constituents are organised and bound together and the larger-scale physical characteristics.

The first three chapters are concerned with the following items:

- basic principles;
- application of principles;
- bulk physical properties.

In this chapter the basic ideas about atoms are developed to explain the different types of bonding which occur. The two main types are *ionic bonding* and *covalent bonding*. An important third type of bonding, *hydrogen bonding*, which only occurs under particular conditions, is also reviewed.

1.1 Atoms, ions and the classification of the elements

All matter, rocks, minerals, water, etc., is made up of atoms. An *element* is an example of matter which contains only one type of atom; for example, diamond contains only carbon atoms, iron contains only iron atoms and gold contains only gold atoms.

A **compound** is an example of matter where two or more elements are chemically combined in some definite fixed ratio; for example, sodium chloride (NaCl) has one sodium for every one chlorine atom. Rocks and minerals are made up of compounds and only the most inert of elements occur pure (in native form) in nature, e.g. gold is found as nuggets.

Atoms are made up of three particles: *protons, neutrons* and *electrons*. Atoms have a small dense nucleus made up of protons and neutrons and is surrounded by the electrons which are very spread out. Most of the mass of an atom is in its nucleus and in addition the nucleus is positively charged. This can be understood by looking at the properties of the particles:

	Relative mass	Relative charge
Proton	1	+1
Neutron	1	0
Electron	1/1836	−1

The hydrogen atom is the simplest of all atoms and contains one proton and one electron. The atom can be drawn as a nucleus with the electron orbiting round it:

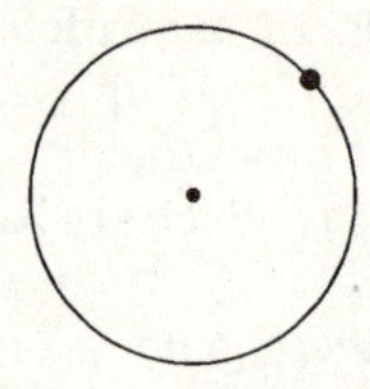

The helium atom has two protons and two neutrons in its nucleus and two electrons orbiting round it:

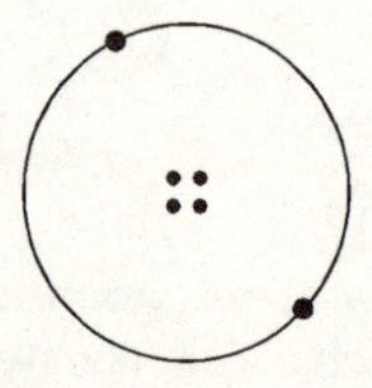

Notice that in neutral atoms the number of positively charged protons is equal to the number of negatively charged electrons. The number of protons is called the *atomic number*. It is 1 for hydrogen and 2 for helium.

The lithium atom has three protons, four neutrons and three electrons:

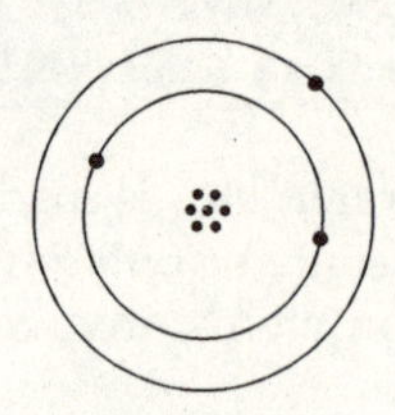

Notice that the third electron is contained in a second shell of electrons outside the first one.

1.1.1 Ions

These are charged. Sodium (Na) has an atomic number of 11 so a neutral atom of sodium contains 11 protons and 11 electrons. If a sodium atom loses one of its electrons it will still contain 11 positively charged protons, but only 10 negatively charged electrons, and so the overall charge will be +1. This is the sodium ion and is written Na^+.

Chlorine (Cl) has a tendency to gain an electron which will give it one extra negative charge. The chlorine ion is written Cl^-.

Ions can come together to form *ionic solids*.

Ionic bonding

In the formation of the ionic solid, sodium chloride, one electron is transferred from a sodium atom to a chlorine atom, so changing the charge of both and resulting in mutual attraction. Sodium chloride can be written Na^+Cl^- to show the charges on the ions which are present, though usually ionic solids are written without the charges being shown in this way. Calcium chloride ($CaCl_2$) is another example of an ionic solid. It contains calcium ions (Ca^{2+}) and two chloride ions (Cl^-) for every calcium ion.

Sodium forms a singly charged *cation* (cation is another way of describing a positive ion) while calcium forms a doubly charged cation. In order to explain this it is necessary to know that electrons in atoms are arranged in shells and sodium has one more electron than it needs to achieve a stable outer shell of eight electrons, while calcium has two more electrons than it needs to achieve a stable outer shell of eight electrons.

Periodic table of the elements and ionic bonding

This is a method of classifying the elements which was first developed in its modern form by Mendeleev (*see* Box 1.1). The formation of ionic bonds and the nature of ions can usefully be discussed in general terms by use of the periodic table. If you look at the periodic table of the elements in Appendix D you will see that group 1 (the first vertical group on the left) contains the elements lithium (Li), sodium (Na), potassium (K), rubidium (Rb) and caesium (Cs). Each of these elements has one more electron than it needs to achieve a stable outer shell of eight electrons, so they form singly charged cations, e.g. Li^+ and chlorides of formula LiCl, NaCl, etc. Similarly for group 2 elements they will form doubly charged cations, e.g. magnesium forms Mg^{2+}, and the chloride has the formula $MgCl_2$.

Negatively charged ions (also called *anions*) occur where the atoms have just less than a stable outer shell of eight electrons, so that they will tend to gain one or more extra electrons in order to have a stable outer shell.

Box 1.1 The periodic table of the elements

The modern form of the periodic table came originally from the Russian scientist, Dimitri Mendeleev. Mendeleev observed that if the elements known at that time were written down in order of increasing atomic weight (now called relative atomic mass), every so often the properties of the elements seemed to repeat in nature; for example, highly reactive light-coloured metals recurred. This *periodic variation* in properties was the basis of what Mendeleev called his *periodic law*. In 1869 Mendeleev proposed that the elements could be written down in rows in order of increasing atomic weight, returning to the start of the next row when properties were repeated. This led to columns of elements (called groups) with similar properties and rows of elements across which trends in properties were observed.

The modern form of the periodic table is shown in Appendix D. We now know that the periodic table is based on writing down the elements in order of increasing atomic numbers (i.e. in order of increasing number of protons and hence of electrons for neutral atoms). The groups (columns of elements) are numbered *across* 1, 2, 3, 4, etc. up to 18 for the rare gases (He, Ne, Ar, Kr, Xe, Rn). Groups 13 to 18 may also be called groups III, IV, V, VI, VII and O, which may be preferable since this represents the number of electrons which may be involved in bonding. The ions formed by some elements of importance in weathering studies are discussed below:

- Sodium (Na) is in group I. It can lose one electron to form Na^+.
- Aluminium (Al) is in group III. It can lose three electrons to form Al^{3+}.
- Chlorine (Cl) is in group VII, so it can gain one electron to give it eight electrons as it forms the chloride ion (Cl^-).

Examples are the elements in group 17 (also called group VII) called the halogens, e.g. fluorine (F), chlorine (Cl), bromine (Br) and iodine (I). They need one more electron and form singly charged anions, e.g. fluorine forms the fluoride ion (F^-). (Note that for anions the name is formed by removing the end of the name from the element and adding the ending 'ide'.)

The elements of group 16 (also called group VI) are two electrons short of the stable eight electrons and will gain two electrons, e.g. oxygen forms the oxide ion (O^{2-}) and sulphur forms the sulphide ion (S^{2-}).

It can be generally observed that ionic solids are formed between metallic elements, which are those occurring to the left of the periodic table, and non-metallic elements, which are those occurring towards the top right-hand corner of the periodic table.

1.1.2 Relative atomic mass (RAM), isotopes and relative molecular mass (RMM)

Atoms of elements have different masses so if we analyse a sample of sodium chloride we would find that for every 23 g of sodium there would be 35.5 g of chloride. Similarly in a weathering process where hydrogen ions displace ions from clay minerals, 1 g of hydrogen ions (H^+) could displace 23 g of sodium ions (Na^+) but could displace 39 g of potassium ions (K^+). Atoms of different elements have different masses because they have different numbers of particles, e.g. a hydrogen atom contains one proton (and one electron). Most of the mass comes from the proton (section 1.1). We can write the hydrogen atom as ^1H where the number 1 shows that hydrogen has a *mass number* of 1.

A helium atom contains two protons and two neutrons (and two electrons) so we say that ^4He has a mass number of 4. Similarly:

- ^{12}C has a mass number of 12 (six protons and six neutrons);
- ^{23}Na has a mass number of 23 (11 protons and 12 neutrons);
- ^{39}K has a mass number of 39 (19 protons and 20 neutrons).

The relative atomic mass (RAM) is defined on the basis of $^{12}C = 12.0000$. So ^1H has a RAM of 1, ^4He has a RAM of 4, ^{23}Na has a RAM of 23 and ^{39}K has a RAM of 39. Many elements have relative atomic masses which are not whole numbers. This arises because elements usually contain a mixture of isotopes.

Isotopes are atoms of the same element (and therefore the same number of protons) but variable numbers of neutrons. For example, hydrogen atoms can exist as:

- ^1H with one proton and no neutrons;
- ^2H with one proton and one neutron;
- ^3H with one proton and two neutrons.

Naturally occurring hydrogen has a relative atomic mass of 1.008 and contains 99.84% of ^1H and 0.016% of ^2H (^2H is also called deuterium). ^3H is radioactive and decomposes, so its concentration in natural hydrogen is vanishingly small (^3H is also called tritium).

If we know the relative amounts of isotopes, we can calculate the relative atomic mass, e.g. for lead (Pb) (*see* Table 1.1).

Isotopes are very important for making measurements of the age of specimens. This is called *isotopic dating* and can be very useful, e.g. for estimating rates of weathering (*see* Chapter 6). An example of the technique is shown in Box 1.2.

From a knowledge of relative atomic masses it is easy to calculate relative molecular mass (RMM) by adding together the relative atomic masses of the elements which make up the compound. For example, for water (H_2O):

$$\begin{aligned} & \quad\;\; \mathrm{H} \qquad\;\; \mathrm{H} \qquad\quad \mathrm{O} \\ \mathrm{RMM} &= 1.008 + 1.008 + 16.000 \\ &= 18.016 \end{aligned}$$

TABLE 1.1 Relative atomic mass of lead

Isotopes	Isotopic mass	Relative amount (%)	Mass of isotopes in 100 atoms of Pb
^{204}Pb	204	2	408
^{206}Pb	206	24	4944
^{207}Pb	207	22	4544
^{208}Pb	208	52	10816
			20722

Average mass of one atom = 20722/100 = 207.22
The RAM of lead is 204.22

and for carbon dioxide (CO_2):

$$\begin{aligned} &\quad\;\; \text{C} \qquad\quad \text{O} \qquad\quad \text{O} \\ \text{RMM} &= 12.000 + 16.000 + 16.000 \\ &= 44.000 \end{aligned}$$

1.1.3 The mole concept

We know that the atoms of different elements have different relative atomic masses and that we can use the values of relative atomic masses to calculate relative molecular masses. We can now use these ideas to calculate amounts of substances involved in a chemical reaction.

To demonstrate the principles we can consider the weathering of limestone by acid rain. For the purposes of simplification we shall assume that acid rain is a solution of sulphuric acid (H_2SO_4) and that limestone is pure calcium carbonate ($CaCO_3$). We can write a chemical equation:

$$CaCO_3 + H_2SO_4 \longrightarrow CaSO_4 + H_2O + CO_2$$

Box 1.2 Isotopic dating using ^{40}K

^{40}K is an isotope of potassium which can capture an electron to become ^{40}Ar:

$$\underset{(19 \text{ protons} + 21 \text{ neutrons})}{^{40}K} + \text{electron} \longrightarrow \underset{(20 \text{ protons} + 20 \text{ neutrons})}{^{40}Ar}$$

The time taken for one-half of the ^{40}K to change to ^{40}Ar is 1.26×10^9 years. This is called the *half-life* $t_{1/2}$.

In order to estimate the age of a sample of rocks they are crushed in a vacuum. The amount of argon present is determined using a mass spectrometer. The amount of argon (^{40}Ar) is compared with the amount of ^{40}K present. The ratio of ^{40}K to ^{40}Ar can then be used to date the rock.

This tells us that one $CaCO_3$ will react with one H_2SO_4 molecule. There are millions of molecules in even tiny amounts of substance, so the idea of molecules reacting together is not very useful when we are considering usual scales of reaction, e.g. grams and kilograms. Therefore we use the mole (abbreviation mol) which is defined by the number of atoms of ^{12}C in 12 grams (0.012 kg) of ^{12}C. This is a very large number of particles, 6.023×10^{23}, called the *Avogadro constant.*

One mole of ^{12}C (6.023×10^{23} atoms) has a mass of 12 g, and 1 mole of hydrogen (6.023×10^{23} atoms) has a mass of 1.008 g because ^{1}H has a mass one-twelfth that of carbon.

To calculate the mass of 1 mole of $CaCO_3$ and of H_2SO_4 we need to know their RMMs:

$$\text{RMM of } CaCO_3 = \underset{40}{Ca} + \underset{12}{C} + \underset{3\times 16}{3O}$$

$$= 100$$

$$\text{RMM of } H_2SO_4 = \underset{2}{2H} + \underset{32}{S} + \underset{64}{4O}$$

$$= 98$$

So 1 mole of $CaCO_3$ has a mass of 100 g and 1 mole of H_2SO_4 has a mass of 98 g.

You can now easily see a pattern, which is that 1 mole of an element is the relative atomic mass expressed in grams; 1 mole of a compound is the relative molecular mass expressed in grams. The term *molar mass* is often used for the mass of 1 mole of an element or a compound.

In our weathering example:

$CaCO_3$	+	H_2SO_4	$\longrightarrow CaSO_4 + CO_2 + H_2O$
1 molecule*		1 molecule	
6.023×10^{23} molecules		6.023×10^{23} molecules	
1 mole (≡100 g)		1 mole (≡98 g)	

(*Note that $CaCO_3$ occurs as an ionic solid, therefore the term 'formula unit' is often used.) So 1 mole of $CaCO_3$ will react with 1 mole of H_2SO_4 and this will produce 1 mole of $CaSO_4$.

In nature we would be dealing with solid limestone ($CaCO_3$) and acid rain as a dilute solution of sulphuric acid. In order to calculate the amount of limestone which could react with the sulphuric acid, we need to know the amount of sulphuric acid. This can be calculated from a knowledge of the volume and concentration which is expressed as mol dm^{-3} (dm = decimetre and is one-tenth of a metre; 1 dm^3 = 1000 cm^3 = 1 litre). Let us assume that the concentration of the acid rain is 0.00005 mol dm^{-3} and that 500 litres (dm^3) falls on a square metre of limestone over a year (this corresponds to an annual rainfall of 50 cm).

Thus 1 dm^3 of 0.00005 mol dm^{-3} of sulphuric acid contains 0.00005 moles, and 500 dm^3 of 0.00005 mol dm^{-3} of sulphuric acid contains $0.00005 \times 500 =$ 0.025 moles of H_2SO_4. If there are 0.025 moles of H_2SO_4 this will react with

0.025 moles of $CaCO_3$. Now, 1 mole of $CaCO_3$ is 100 g (the RMM expressed in grams). So 0.025 moles of $CaCO_3$ is $100 \times 0.025 = 2.5$ g of $CaCO_3$. This is the weight of limestone which would react if all of the sulphuric acid in rain reacted with the limestone.

1.1.4 Equivalents (and milliequivalents)

In modern applications the mole has replaced the *equivalent* as a measure of amount of substance. The milliequivalent (meq) is 1/1000 of an equivalent. However, the equivalent is still occasionally used in text books and papers in the Earth sciences and so you need to understand it.

The mole was described in the previous section. You saw that 1 mole of a substance has a mass equal to its relative molecular mass (RMM) expressed in grams. One equivalent of a substance is its equivalent weight (EW) expressed in grams. The EW of a substance is the amount which will react completely with the equivalent weight of another substance. EW is equal to or is a fraction of RMM.

We can show how this works by taking two examples:

- 1 mole of HCl has a mass of 36.5 g (RMM = 36.5)
  1 equivalent of HCl has a mass of 36.5 g (EW = 36.5)
  The EW is equal to the RMM.
- 1 mole of H_2SO_4 has a mass of 98 g (RMM = 98)
  1 equivalent of H_2SO_4 has a mass of 49 g (EW = 49)
  The EW is equal to half the RMM.

This is because H_2SO_4 has two replaceable hydrogen atoms, whereas HCl has only one replaceable hydrogen atom.

Now consider sodium hydroxide (NaOH). Its RMM is 40 and its EW is 40 since it only has one OH group. The reaction of 1 mole of sodium hydroxide with the two acids described above is

$$\underset{\substack{\text{1 mole}\\ \text{40 g}\\ \text{1 equivalent}}}{\text{NaOH}} + \underset{\substack{\text{1 mole}\\ \text{36.5 g}\\ \text{1 equivalent}}}{\text{HCl}} \longrightarrow \text{NaCl} + \text{H}_2\text{O}$$

$$\underset{\substack{\text{1 mole}\\ \text{40 g}\\ \text{1 equivalent}}}{\text{2NaOH}} + \underset{\substack{\text{0.5 mole}\\ \text{49 g}\\ \text{1 equivalent}}}{\text{H}_2\text{SO}_4} \longrightarrow \text{Na}_2\text{SO}_4 + 2\text{H}_2\text{O}$$

In both cases, 1 equivalent reacts with 1 equivalent, and this was the advantage of this system since in terms of moles in the second case you need half the number of moles of H_2SO_4 compared to the number of moles of NaOH. Now consider $Ca(OH)_2$. The RMM is 74; what is its EW? It contains two OH^- groups, therefore its $EW = 74/2 = 37$. Its reaction with the acids is

$$\underset{\substack{0.5\text{ mole}\\1\text{ equivalent}}}{Ca(OH)_2} + \underset{\substack{1\text{ mole}\\1\text{ equivalent}}}{2HCl} \longrightarrow CaCl_2 + 2H_2O$$

$$\underset{\substack{1\text{ mole}\\1\text{ equivalent}}}{Ca(OH)_2} + \underset{\substack{1\text{ mole}\\1\text{ equivalent}}}{H_2SO_4} \longrightarrow CaSO_4 + 2H_2O$$

Again, in both cases 1 equivalent reacts with 1 equivalent, but in the case of HCl we need only 0.5 mole of the $Ca(OH)_2$.

This system can be extended to ions such as Ca^{2+}, Mg^{2+} and Al^{3+} where they are considered to replace H^+. Since Ca^{2+} would displace $2H^+$:

$$1\text{ millequivalent of }Ca^{2+} = 0.5\text{ millimoles of }Ca^{2+}$$

and for Al^{3+}:

$$1\text{ milliequivalent of }Al^{3+} = 0.333\text{ millimoles of }Al^{3+}$$

These calculations were shown in milliequivalents to illustrate that quantities involved in weathering, for example of rocks, would often be expressed as milliequivalents per gram of rock. Another example of the traditional use of 'equivalents' is to be found in calculations made for the 'liming of soils'.

In this section we have only discussed acids and bases. Equivalent weight can also be used for reduction and oxidation where

- the EW of a reducing agent $= \dfrac{\text{RMM}}{\text{number of electrons lost}}$
- the EW of an oxidising agent $= \dfrac{\text{RMM}}{\text{number of electrons gained}}$

1.1.5 The nature of ionic bonding

We have seen that in the formation of an ionic solid, e.g. NaCl, there is a transfer of electrons from the atom which becomes the cation, e.g. Na forms Na^+, to the atom which becomes the anion, e.g. Cl forms Cl^-. These ions are attracted to each other and we need to study the way in which the ions are packed together to form the solid. You should think of the ions as hard spheres packed together as closely as possible as this maximises the attractions between them.

Packing of spheres

Let us start by considering how spheres pack in a single layer. Two possible arrangements are shown in Fig. 1.1. Of the two arrangements (a) and (b), the second gives the maximum occupancy of the available space and is said to be *close packing of spheres*.

Crystals have a three-dimensional structure, so we must consider the arrangements which occur when there is more than one layer of spheres;

FIGURE 1.1 The packing of a single layer of spheres.

this is shown in Fig. 1.2. If we call the first layer of spheres A, then there is a set of hollows on the top of A in which we can place the second layer of spheres called B. This is shown as arrangement (a) in the figure. The third layer can now be made by placing spheres in the hollows on the second layer. There are two ways of arranging the third layer while retaining close packing. The third layer may be placed in hollows so as to be directly above layer A. Alternatively the third layer may be C, where it is created by placing spheres in hollows which are not directly above spheres in layer A. So two forms of close packing in three dimensions can occur. These are ABCABC, etc., which is called *cubic close-packing* and ABAB, etc., which is called *hexagonal close-packing*. Both of these forms of close packing give 74 per cent of available space occupied.

Another important aspect of the close packing of spheres is the holes (also called sites or interstices) which are created between the close-packed spheres. There are two types of hole and they are called *tetrahedral sites* and *octahedral sites*. When a sphere is placed on three others a tetrahedral arrangement of the spheres is created, and the hole at the centre of four spheres is called a tetrahedral site (*see* Fig. 1.3). The octahedral site arises

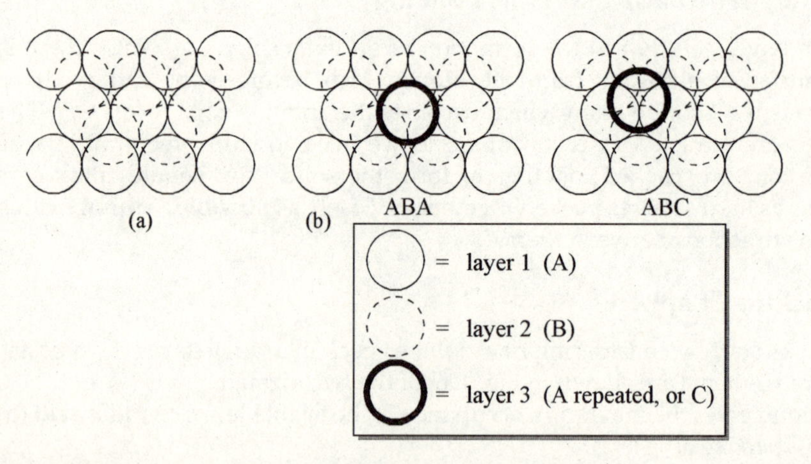

FIGURE 1.2 Three-dimensional packing of spheres. (a) Two layers of close-packed spheres and (b) three layers of close-packed spheres.

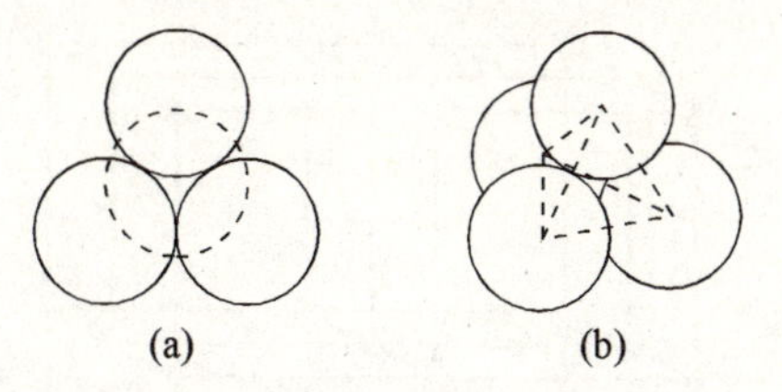

FIGURE 1.3 Tetrahedral sites in close-packed spheres. (a) View from above and (b) three-dimensional view.

between two layers of close-packed spheres and is created from the unoccupied hollows between the two layers of close-packed spheres (*see* Fig. 1.4).

We can understand how this arises by thinking about how layer B is created by placing spheres in hollows on the top of layer A. Only half of the available hollows are used, and this leaves a set of unoccupied hollows which are the octahedral sites.

It is possible to fit small spheres into tetrahedral sites or into octahedral sites. If we call the radius of the small sphere r_1 and the radius of the larger close-packed spheres r_2, then

for a tetrahedral site $r_1/r_2 = 0.225$

for an octahedral site $r_1/r_2 = 0.414$

These values are called the radius ratio values for the different types of site and become important when we consider how the packing of spheres relates to the structure of ionic solids. Notice that octahedral sites can hold larger spheres than tetrahedral sites.

Sodium chloride

In a simple ionic solid like sodium chloride there are relatively large chloride ions and relatively small sodium ions. The model of close packing of spheres

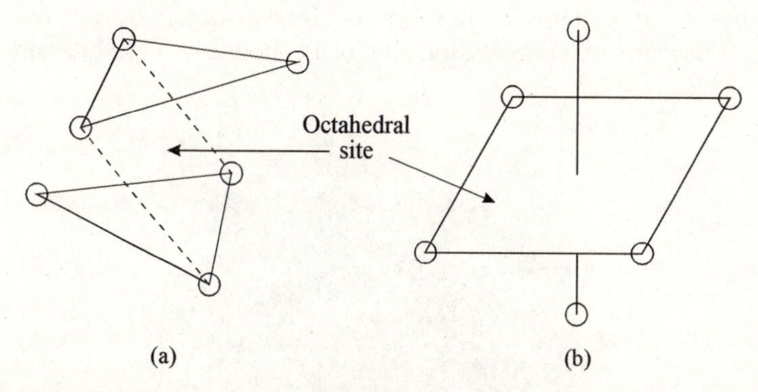

FIGURE 1.4 Octahedral sites in close-packed spheres. (a) Shows how an octahedral site arises between two layers of close-packed spheres. (b) How an octahedral site is shown in some text books.

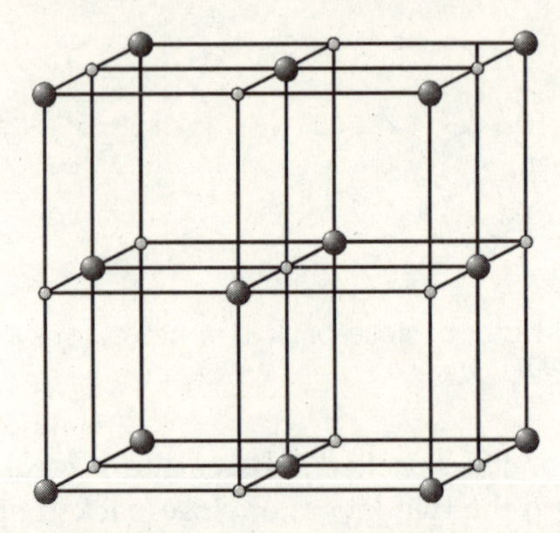

FIGURE 1.5 An exploded diagram of the structure of sodium chloride. Note that the ions are in equivalent positions, but the larger spheres may be taken to represent the chloride ions which are larger.

can now be used to describe the structure of sodium chloride: The chloride ions are in a close-packed type of arrangement with sodium ions in all of the available octahedral sites in the structure.

An important thing to note about this statement is that the chloride ions are said to be a cubic close-packed *type* of arrangement. The chloride ions are not actually close packed because the sodium ions are too large to fit exactly in the octahedral sites. The radius ratio for a small sphere to fit exactly in the octahedral site is 0.414. The ionic radius of Na^+ is 0.098 nm (note 1 nanometre (nm) = 10^{-9} metres) while the ionic radius of Cl^- is 0.181 nm so:

$$r_1/r_2 = 0.098/0.181 = 0.54$$

The radius ratio value of 0.54 is greater than the value of 0.414 required for the sodium ion to fit exactly in the octahedral site, so the chloride ions are not in contact with each other as they would be in a truly close-packed structure.

The structure of sodium chloride can be shown in various ways. In Fig. 1.5 the relative arrangements of the ions is clearly shown by the use of an exploded diagram. A space filling model is shown in Fig. 1.6, and while

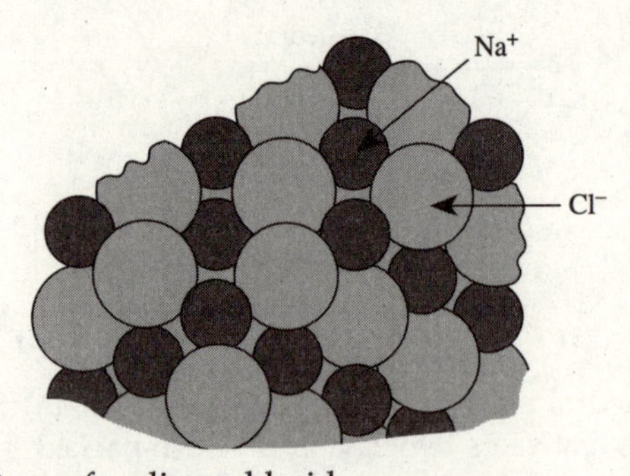

FIGURE 1.6 The structure of sodium chloride.

this is closer to reality in that the ions are shown in contact, the relative arrangements are much less clear.

In the structure of sodium chloride all of the available octahedral sites are filled, and since for every sphere in a close-packed structure there is one octahedral site, this ensures that sodium chloride has equal numbers of sodium and chloride ions as expected from the formula NaCl. With regard to tetrahedral sites there are two for every sphere, and this will be of importance when we consider other structures later on.

The diagram shown in Fig. 1.5 demonstrates a number of other important features:

- The arrangements of both ○ ions and ● ions are face-centred cubic showing that the ions are in equivalent positions in the structure. Face-centred cubic is an arrangement found in cubic close packing.
- It is found that each ○ ion is surrounded by six ● ions. We say that the *coordination number* of sodium ions is 6 and the *coordination number* of chloride ions is 6. (*The coordination number of an ion is defined to be the number of its nearest neighbours.*)
- The arrangements of ○ and ● ions are seen to be equivalent and thus the original description of NaCl could have been 'the sodium ions are in a cubic close-packed type of arrangement with the chloride ions in all of the available octahedral sites in the structure'. The earlier description is usually preferred as the chloride ions are larger than the sodium ions.

Unit cell

The description of ionic solids relies upon identifying a small unit which is representative of the whole structure. This is called the *unit cell*. The unit cell can be thought of as being a building brick and, just as bricks are stacked up to build a wall, so unit cells are stacked to give the structure of the ionic solid. This means that if an ion is on the face of a unit cell it is regarded as being cut in half. (Of course it is not actually cut in half – we merely think of it in this way!) See Fig. 1.7 for the unit cell of sodium chloride. The ion on the face is shared between two unit cells, which is why it is cut in half. For the purpose of counting up ions in the unit cell which we are considering, an ion on one

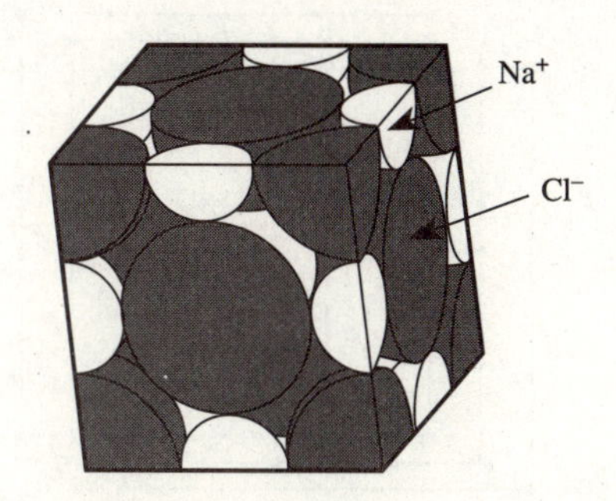

FIGURE 1.7 The unit cell of sodium chloride.

TABLE 1.2 Counting ions in a unit cell

Position of ion	Count in the unit cell
Within the cell	1
On a face	$\frac{1}{2}$
On an edge	$\frac{1}{4}$
On a corner	$\frac{1}{8}$

face counts one half. Ions on the edges of a unit cell are shared between four unit cells and so only count for one-quarter in the unit cell. Similarly ions on the corners of unit cells are shared between eight unit cells and count one-eighth in the unit cell. This is summarised in Table 1.2. The number of ions in the unit cell for sodium chloride can now be counted up using the diagram in Fig. 1.5. For ○ there are:

12 on edges	$12 \times \frac{1}{4}$	$= 3$
1 within the cell	1×1	$= 1$
Total		4

For ● there are:

8 on corners	$8 \times \frac{1}{8}$	$= 1$
6 on face centres	$6 \times \frac{1}{2}$	$= 3$
Total		4

There are four sodium ions and four chloride ions in the unit cell of sodium chloride.

Other ionic solids

The sphere model can now be extended to other ionic solids. The compound zinc sulphide (ZnS) is interesting as it occurs naturally in two crystalline forms called *zinc blende* and *wurtzite*. The structures are shown in Figs 1.8 and 1.9 respectively. Zinc blende has a cubic close-packed type of

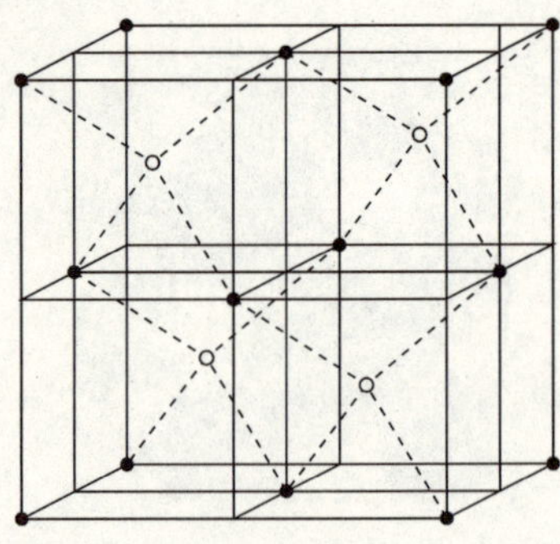

FIGURE 1.8 The structure of zinc blende (ZnS).

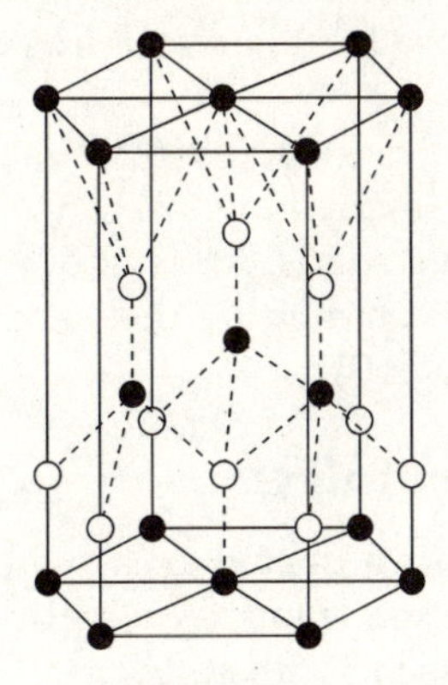

FIGURE 1.9 The structure of wurtzite (ZnS).

arrangement of sulphide ions (S^{2-}) with zinc ions (Zn^{2+}) in half of the available tetrahedral sites. Only half of the available sites are occupied because for every sphere in a close packed structure there are two tetrahedral sites and, since only one Zn^{2+} is required for one S^{2-} to give the formula ZnS, only half of the tetrahedral sites are occupied. Wurtzite differs from zinc blende only in that it has a hexagonal close-packed type of arrangement of S^{2-} rather than a cubic close-packed type of arrangement.

The structure of other important minerals can be described; for example, fluorite (CaF_2) has a cubic close-packed type of arrangement of Ca^{2+} with F^- in all of the available tetrahedral sites as shown in Fig. 1.10.

For the mineral ilmenite ($FeTiO_3$) it is found that there is a hexagonal close-packed type of arrangement of oxide ions (O^{2-}) with iron(II) ions

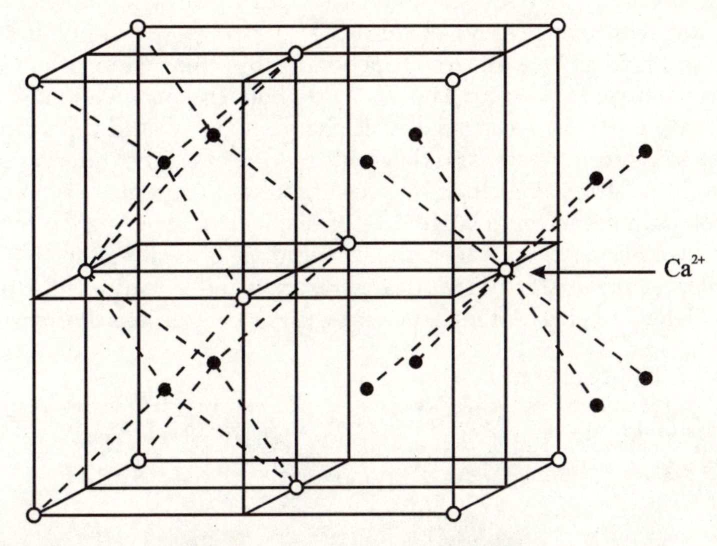

FIGURE 1.10 Structure of fluorite, calcium fluoride (CaF_2).

(Fe^{2+}) and titanium(IV) ions (Ti^{4+}) in two-thirds of the available octahedral sites.

The sphere packing model is useful in visualising how ionic solids are formed and can be used to describe a whole range of simple and complicated structures. Apart from ionic solids, ideas about packing of spheres are of crucial importance in a discussion of the behaviour of gravel, silt, etc., exposed to water (*see* Chapter 3).

Lattice energy (Lattice enthalpy)

Positive and negative ions are attracted to each other, and it is these forces of attraction which hold ionic solids together. Lattice energy can be defined as 'the amount of energy which is liberated when 1 mole of ionic solid is formed from gaseous ions coming together from infinity' (the mole was discussed in section 1.1.3). For sodium chloride 1 mole is 58.5 g and the lattice energy is the energy given out (liberated) when 1 mole (23 g) of gaseous sodium ions (Na^+) and 1 mole (35.5 g) of chloride ions (Cl^-) come together to form 1 mole (58.5 g) of solid sodium chloride. This can be written:

$$Na^+(g) + Cl^-(g) \longrightarrow NaCl(s)$$

The values of lattice enthalpy are expressed as kilojoules (kJ) per mole (mol) ($kJ\,mol^{-1}$). They are given as negative quantities, as heat is given out in the process of forming the ionic solid. The larger the negative value, the greater is the binding together of the ionic solid. By convention the term 'higher lattice enthalpy' is used to indicate a larger negative value of lattice enthalpy. Some values are given in Table 1.3. The values for the different compounds differ considerably. How can this be explained?

In sodium chloride (NaCl) the ions are Na^+ and Cl^- while in calcium chloride ($CaCl_2$) they are Ca^{2+} and Cl^-. In calcium chloride the presence of an ion with two positive charges (Ca^{2+}) as opposed to one unit charge (Na^+) in NaCl gives a greater attraction for the chloride ions, so the lattice energy is greater. Now for calcium oxide both the positive ion (Ca^{2+}) and the negative ion (O^{2-}) carry a double charge, so we would expect the lattice energy to be even greater and this is indeed the case. The differing ability of ions to have strong ionic bonding has important consequences in weathering, for example in the relative ease of displacement of one ion by another in the weathering of soils by aqueous solutions of ionic solids.

Other factors apart from the charge on an ion affect lattice enthalpy. Small ions will tend to give a larger lattice enthalpy than large ions. We can illustrate

TABLE 1.3 Lattice energies

Compound	Lattice energy ($kJ\,mol^{-1}$)
Sodium chloride (NaCl)	−781
Calcium chloride ($CaCl_2$)	−2197
Calcium oxide (CaO)	−3523

TABLE 1.4 Lattice energies of the halides of the alkali metals (kJ mol^{-1})

Alkali metal ion	Fluoride F^-	Chloride Cl^-	Bromide Br^-	Iodide I^-
Li^+	−1029	−849	−804	−753
Na^+	−915	−781	−743	−699
K^+	−813	−710	−679	−643
Rb^+	−779	−685	−656	−624

TABLE 1.5 Ionic radii of the alkali metal ions and halide ions

Alkali metal ion	Ionic radius (nm)	Halide ion	Ionic radius (nm)
Li^+	0.068	F^-	0.133
Na^+	0.098	Cl^-	0.181
K^+	0.133	Br^-	0.196
Rb^+	0.148	I^-	0.219

this by comparing sodium chloride (NaCl) and potassium chloride (KCl). Sodium ions (Na^+) and potassium ions (K^+) have the same charge, but Na^+ has an ionic radius of 0.098 nm while K^+ has an ionic radius of 0.133 nm. The single charge of the sodium ion is therefore spread over a smaller surface area than the single charge of the potassium ion. The greater charge intensity of the sodium means that the chloride ions are attracted more strongly to the sodium ion. The lattice enthalpy of sodium chloride is −781 kJ mol^{-1} while potassium chloride is held together more weakly and has a lower lattice enthalpy of −710 kJ mol^{-1}. These values are shown in Table 1.4.

In general there is a good correlation between the size of ions and the lattice energies of the alkali metal halides. The ionic radii are given in Table 1.5. As the size of the alkali metal ion increases, the lattice enthalpy decreases in any of the series of fluorides or chlorides or bromides or iodides. Similarly if you look at one line in Table 1.4 for any particular metal ion, the lattice enthalpy changes in the order

$$F^- > Cl^- > Br^- > I^-$$

which parallels the change in ionic radius:

$$F^- < Cl^- < Br^- < I^-$$

Often in weathering processes we are not concerned with the single ionic solids described in this section; however, the principles of the effects of ion charge and size are often useful even in more complicated systems.

1.2 COVALENT BONDING

In ionic bonding electrons are transferred from one atom to another, but in covalent bonding pairs of electrons are *shared* between atoms. The simplest

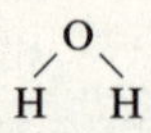

FIGURE 1.11 The water molecule (angular).

example is the dihydrogen molecule (two hydrogens, H_2). A hydrogen atom has only one electron and when two hydrogen atoms combine the molecule is formed and can be represented as:

$$H:H \quad \text{or} \quad H{-}H$$

Other examples of molecules are:

Chlorine	Cl : Cl	or	Cl–Cl
Hydrogen chloride	H : Cl	or	H–Cl

With hydrogen and chlorine atoms, each atom provides only one electron for sharing, but for oxygen atoms two electrons are provided for sharing, so that an oxygen atom combines with two atoms of hydrogen to form the water molecule (H_2O). The angular water molecule is shown in Fig. 1.11.

In O_2 a double bond is formed:

$$O{=}O$$

Double bonds are found in a few other molecules, e.g. carbon dioxide (CO_2):

$$O{=}C{=}O$$

In dinitrogen (N_2) a triple bond occurs as the atoms each have three electrons to share:

$$N{\equiv}N$$

In molecules with only two atoms (called diatomic molecules) there is no possibility for variation of shape. In molecules with three atoms (called triatomic molecules) or more, various structural arrangements are possible; for example, you may have noticed earlier that the water molecule is angular while carbon dioxide is linear. This variation in behaviour can be explained by the valence shell electron pair repulsion theory (VSEPR) which is explained in Box 1.3.

It is not essential to have a grasp of this theory, but the recognition of the different shapes of some molecules is important, particularly because molecular shape and physical properties may be linked. An important example is the angular shape of the water molecule which ensures that it is an excellent solvent for many substances which are affected by weathering (*see* Chapter 2).

We have seen that covalent bonding occurs in simple molecules, but it may also occur in complex ionic species, especially oxo-anions. An oxo-anion is so named because it contains oxygen, hence oxo-, and anion because it is a negatively charged ion. Calcium carbonate ($CaCO_3$) contains calcium ions (Ca^{2+}) and carbonate ions (CO_3^{2-}). The carbonate ion is a flat species described as trigonal planar (*see* Fig. 1.12). The sulphate ion (SO_4^{2-}) has

Box 1.3 Valence shell electron pair repulsion theory (VSEPR)

This is concerned with the shape of molecules and is based on the idea that pairs of electrons will be as far apart as possible so as to minimise the repulsions between them. The arrangement are as shown below.

Pairs of electrons	Arrangement
2	Linear
3	Trigonal planar 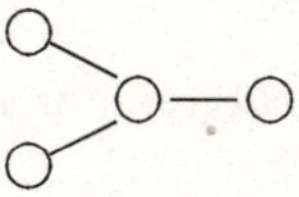
4	Tetrahedral 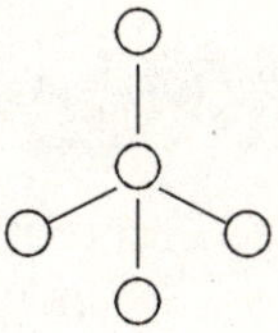
5	Trigonal bipyramid 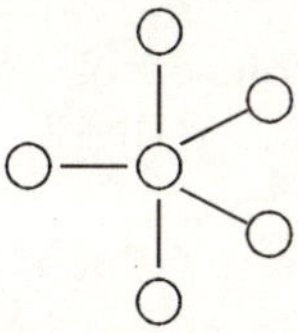
6	Octahedral 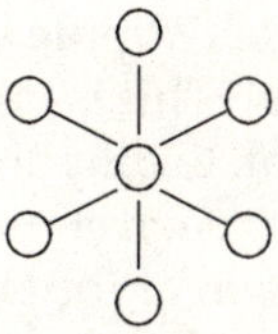

We shall explain how it works for water but the theory is of good general applicability. In water (H_2O) the oxygen has six outermost electrons (electrons in the valence shell); each hydrogen has one outermost electron:

O	6 electrons
2 × H	2 electrons
Total	8 electrons
therefore	4 pairs of electrons

These are arranged tetrahedrally and there are two bond pairs of electrons (O bonded to H atoms) and two lone pairs of electrons. When we look at the water molecule we see an angular H–O–H system.

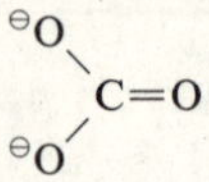

$$^{\ominus}O \backslash C{=}O \,/\, ^{\ominus}O$$

FIGURE 1.12 The carbonate ion (CO_3^{2-}).

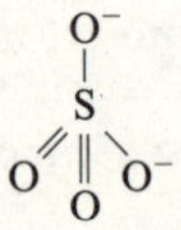

$$O^- - S(=O)(=O) - O^-$$

FIGURE 1.13 The sulphate ion (SO_4^{2-}) (tetrahedral).

four oxygen atoms arranged tetrahedrally around a central sulphur atom (*see* Fig. 1.13).

1.3 INTERMOLECULAR FORCES (VAN DER WAALS FORCES)

Within a molecule, atoms are held together by the sharing of pairs of electrons. *Between* molecules there are three attractive forces which can be identified:

- dipole–dipole forces;
- induced dipole–induced dipole forces (also called London or dispersion forces);
- dipole–induced dipole forces.

These three forces are sometimes collectively called *van der Waals forces.*

Dipole–dipole forces can only occur if a molecule has a permanent dipole moment. This is present if the distribution of electrons in an atom is not symmetrical. In dihydrogen (H_2) the pair of electrons are equally shared between the hydrogen atoms. In hydrogen chloride (HCl) the chlorine has a stronger attraction for the shared pair of electrons than the hydrogen. This makes the chlorine slightly negative and the hydrogen slightly positive. This can be written:

$$^{\delta+}H{-}Cl^{\delta-}$$

(The Greek symbol δ (delta) is often used to indicate a small amount.) The polar nature of the HCl bond means that HCl has a dipole moment of 1.04 D (the debye unit (symbol D) has the value 1 for a unit positive charge of a proton separated by 100 pm (pm $= 10^{-12}$ metres) from an equal and opposite charge).

It is important to distinguish bond polarity and dipole moment, e.g. CO_2 has polar C=O bonds, but the linear structure of the molecule means that the polar C=O bonds cancel each other out; overall the distribution of electrons is symmetric and so the dipole moment of CO_2 is zero. In contrast to CO_2, water has an angular structure and the presence of the polar O–H bond

TABLE 1.6 Melting points and relative molecular masses of the halogens

Halogen	Melting point (°C)	Relative molecular mass
Fluorine (F_2)	−220	39
Chlorine (Cl_2)	−101	71
Bromine (Br_2)	−7	160
Iodine (I_2)	114	254

results in a dipole moment of 1.8 D. This is discussed in more detail in section 4.2.

Molecules which have a non-zero dipole moment are attracted to each other because of the attraction of the positive end of one molecule to the negative end of another molecule.

Induced dipole–induced dipole forces arise because at any one time in a molecule the electrons may be unequally distributed, so that a temporary dipole moment occurs. The δ^+ in the molecule will tend to line up with the δ^- of an adjacent molecule giving an attractive force. This can occur with both atoms and molecules and becomes stronger as the electron clouds of atoms or molecules become larger or more polarisable. This is shown by the melting points of the halogens which increase as the relative molecular masses and hence the electron clouds become larger (*see* Table 1.6).

The **dipole–induced dipole force** is small relative to the other two types and can be ignored.

Intermolecular forces play an important role in some weathering processes; for example, the effectiveness of water as a solvent is influenced by the intermolecular forces in the liquid. The weak dispersion forces in some minerals such as micas gives rise to a flaky nature because of cleavage planes where the ions are only held together by van der Waals forces.

1.4 HYDROGEN BONDING

As the name implies, this type of bonding involves hydrogen when it is combined with certain elements. The significant features for weathering are

- how hydrogen bonding influences the properties of water;
- the occurrence of hydrogen bonding in some silicate minerals and particularly clays (*see* Chapter 10), as well as hydroxide minerals.

The occurrence of hydrogen bonding in water can be demonstrated by observing the boiling points of the hydrides (combinations of hydrogen and another element) of group VI elements, which are oxygen (O), sulphur (S), selenium (Se) and tellurium (Te). The hydrides are H_2O, H_2S, H_2Se and H_2Te. A plot of the boiling point of these hydrides against the period of the periodic table is shown in Fig. 1.14.

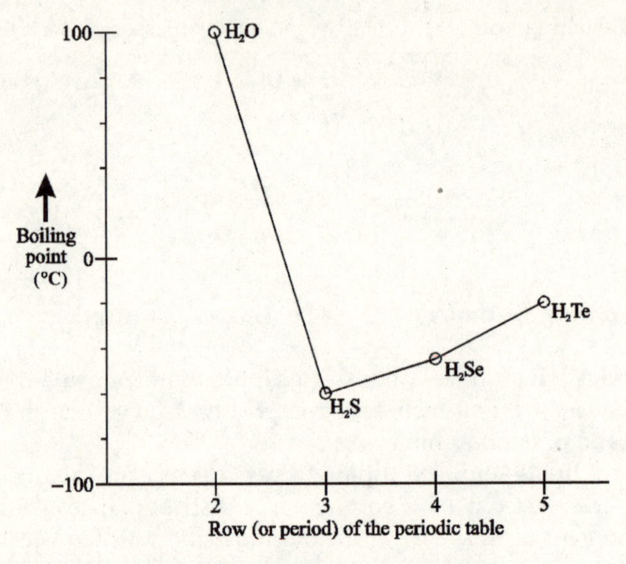

FIGURE 1.14 The boiling points of the hydrides of group VI.

Water is plainly out of line with the other hydrides of the group. For H_2S to H_2Se to H_2Te the boiling point increases smoothly, as the dispersion forces (section 1.3) increase with increases in relative molecular mass. For water there is obviously some extra force holding the molecules together, and so raising the boiling point compared to the other compounds. The extra force is *hydrogen bonding*.

Hydrogen shows this important property because the hydrogen atom has a very simple structure consisting of one proton as its nucleus and one electron. Hydrogen bonding occurs when the hydrogen atom is covalently bonded to another atom with strong electron-attracting powers, such as oxygen. The oxygen attracts the pair of electrons in the O–H bond towards itself and this leaves the back of the hydrogen atom, and hence its positive nucleus, exposed. The positive 'back side' of the hydrogen atom is attracted to a lone pair of electrons on an oxygen atom in an adjacent water molecule. This attraction is hydrogen bonding. Figure 1.15 shows hydrogen bonding between two water molecules, but hydrogen bonding actually occurs between a large number of water molecules.

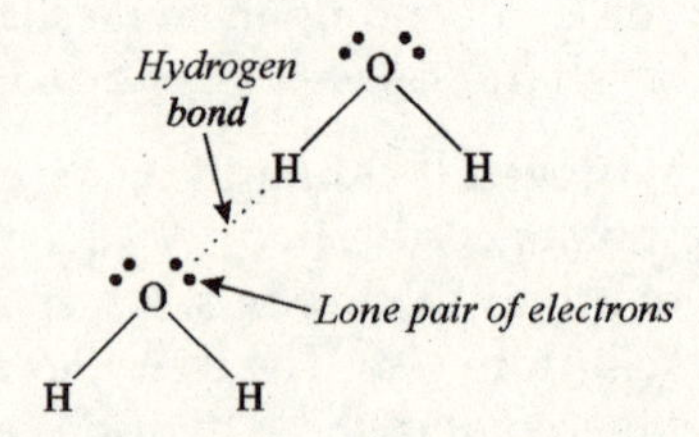

FIGURE 1.15 Hydrogen bonding between two water molecules.

Readings for further study

There are numerous general texts written on 'science' or 'general chemistry' for first- and second-year undergraduate college students, particularly directed at the USA market. These are often excellent teaching texts which are superbly presented. Another good source for the work in this chapter is basic inorganic chemistry text books.

Atkins, P. and Jones, L. 1997: *Chemistry, molecules, matter and change,* 3rd edn. New York: W.H. Freeman. This book is the latest in a variety of books involving authorship by P.W. Atkins. An earlier also excellent book was *General chemistry*. The text is rich with examples and contains clear explanations supported by self-assessment exercises.

Cotton, F.A., Wilkinson, G. and Gauss, P.L. 1994: *Basic inorganic chemistry,* 3rd edn. New York: Wiley. This widely used text book contains clear explanations of basic aspects of atoms and how they bond together.

2

APPLICATION OF PRINCIPLES: STRUCTURES OF SILICATES AND OTHER EARTH MATERIALS

In this chapter the principles of bonding developed in Chapter 1 are used to explain the structures of some important earth materials. The structures of the silicates in particular are quite involved, and include ionic and covalent bonding as well as in some cases hydrogen bonding. The basic structural features developed in this chapter are built on later in the book; for example, to discuss hydration in Chapter 7, and silicates as products of weathering and iron oxide minerals as products of weathering in Chapter 10.

The Earth's crust varies in thickness between 10 and 35 km. The crustal abundance of the 20 most common elements is shown in Table 2.1. Oxygen and silicon taken together account for nearly three-quarters of the mass of the crust, and this demonstrates their importance and why silicates take up most of this chapter. Other important groups of compounds are dealt with at the end of the chapter – carbonates, iron oxides and compounds of aluminium - as typical examples of ionic solids which were described in Chapter 1.

2.1 SILICATES

Silica and silicate minerals are the most common materials in the Earth's crust. They occur not only as components of rocks but also as weathered products, e.g. clays, soils and sands. The silicate minerals also include the aluminosilicates in which aluminium replaces some of the silicon in the silicate structure. Relatively few structural types are of major weathering importance and these will all be introduced in this chapter. The basis of all of the silicates is a simple structural unit which is the SiO_4 tetrahedron,

TABLE 2.1 The percentage by weight of the 20 most abundant elements in the Earth's crust

Element	Symbol	wt%
Oxygen	O	46.67
Silicon	Si	27.72
Aluminium	Al	8.13
Iron	Fe	5.00
Calcium	Ca	3.60
Sodium	Na	2.83
Potassium	K	2.59
Magnesium	Mg	2.09
Titanium	Ti	0.44
Hydrogen	H	0.14
Phosphorus	P	0.12
Manganese	Mn	0.10
Fluorine	F	0.08
Sulphur	S	0.05
Chlorine	Cl	0.05
Barium	Ba	0.04
Carbon	C	0.03
Rubidium	Rb	0.03
Zirconium	Zr	0.02
Chromium	Cr	0.02

in which a silicon atom is surrounded by a tetrahedral arrangement of four oxygen atoms (*see* Fig. 2.1).

The most simple of all of the structures involving the SiO_4 tetrahedron is silicon dioxide (SiO_2), which is also called *silica* (*see* Box 2.1). The common form of this compound is quartz, which is a hard solid commonly found as beach sand. In silica the SiO_4 tetrahedra (the plural of tetrahedron) share all four of their oxygen atoms with each other (*see* Fig. 2.2).

This arrangement of silicon and oxygen gives a very rigid structure held together with strong covalent Si–O bonds. In the silicates a range of structures with varying properties occurs because the SiO_4 tetrahedra can join together by sharing oxygen atoms to form:

- chains;
- double chains (ribbons);

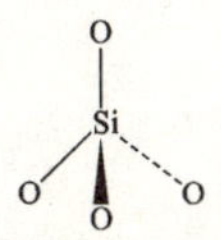

FIGURE 2.1 The SiO_4 tetrahedron found in silicate structures.

Box 2.1 Silica – naming and formula

Silicon dioxide is usually called silica. The ending 'a' added to the shortened name of the element is a classical way of naming oxides; for example, the elements magnesium (Mg) and aluminium (Al) both form compounds which are of importance to weathering. Magnesium can occur in magnesium oxide (MgO), also called *magnesia*, while aluminium can occur in aluminium oxide (Al_2O_3), also called *alumina*.

The modern systematic method of naming these oxides would be to call silicon dioxide (SiO_2) silicon(IV) oxide. The oxidation number of the silicon 'IV' tells you that the charge on the silicon is +4, so there must be two doubly charged oxide ions (O^{2-}) to make the compound electrically neutral and therefore the formula is SiO_2.

Magnesium oxide would be written magnesium(II) oxide, so there needs to be one doubly charged oxide ion (O^{2-}) to make magnesium oxide electrically neutral, and therefore the formula is MgO.

- rings;
- sheets;
- three-dimensional (3-D) networks.

These structures are given particular names as shown in Table 2.2. The negatively charged silicate structures are counterbalanced by positive ions which can fit into cavities in between the silicate anions. The aluminium ion (Al^{3+}) is of particular importance because it has the ability to replace silicon at the centre of SiO_4 tetrahedra, as well as to act like most positive ions in fitting into cavities within the silicate structures. The silicates will now be considered on the basis of an increasing number of oxygen atoms shared between tetrahedra, starting with zero.

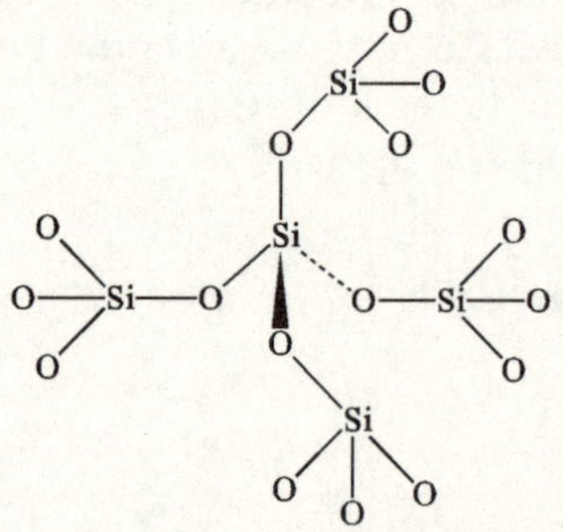

FIGURE 2.2 Part of the structure of SiO_2 showing the sharing of all four oxygen atoms of SiO_4.

TABLE 2.2 Types of silicates

Name	Structural group	General formula	Number of oxygen atoms shared
Nesosilicates ('island') (also called *Orthosilicates*)	Tetrahedra	SiO_4^{4-}	0
Sorosilicates	Double tetrahedra	$Si_2O_7^{2-}$	1
Cyclosilicates	Closed rings of tetrahedra	$(SiO_3^{2-})_n$ $n = 3, 4, 6$	2
Inosilicates	(1) Chains of tetrahedra (2) Double chains of tetrahedra ('ribbons')	$(SiO_3^{2-})_n$ $(Si_4O_{11}^{2-})_n$	2; 2 and 3
Phyllosilicates	Sheets of tetrahedra	$(Si_2O_5^{2-})_n$	3
Tectosilicates	3-D framework of tetrahedra	SiO_2	4

2.1.1 No shared oxygen atoms

In this case the unit is SiO_4^{4-}. The reason for the −4 charge is that the silicon has a +4 charge (Si^{4+}), and each oxygen has a −2 charge (O^{2-}) so that:

$$\begin{array}{lll} \text{Si} & 4\text{O} & \\ +4 & -8 & \text{so charge} = -4 \end{array}$$

The SiO_4^{4-} tetrahedra pack together like typical ionic solids (*see* Chapter 1), and this creates sites in which positive ions are present. Compounds which contain SiO_4^{4-} are called *orthosilicates*. A simple example is zircon, $ZrSiO_4$. The zirconium ion has a +4 charge (Zr^{4+}) to counterbalance the −4 charge of the SiO_4^{4-} unit. Another example is garnet, $Ca_3Al_2(SiO_4)_3$. We can check to see that this is electrically neutral:

$$\text{Ca} = +2 \quad (Ca^{2+})$$

$$\text{Al} = +3 \quad (Al^{3+})$$

So we have

$$\begin{array}{rr} Ca_3 = & +6 \\ Al_2 = & \underline{+6} \\ \text{Total} & +12 \end{array}$$

which counterbalances the $3 \times -4 = -12$ of the three SiO_4^{4-} units.

These materials are typical ionic solids being hard and brittle. The *olivine* group of minerals contains the SiO_4^{4-} unit with divalent cations, e.g. Mg^{2+}, Fe^{2+}, Mn^{2+} and Ca^{2+}, and has the general formula M_2SiO_4 (M = metal ion). The olivines are one of the most easily weathered groups of fine-grained materials in soils owing to their simple structure and low lattice energy. The danger of generalising about ease of weathering is shown by the fact that zircon ($ZrSiO_4$), which contains the Zr^{4+} ion and has a higher lattice energy, is one of the most inert of all the silicates. In fact zircon is used as a standard against which the weathering rate of other silicates is judged (Chapter 9).

2.1.2 One shared oxygen atom

These are again typical ionic solids, but the anion is slightly more complicated as it consists of $Si_2O_7^{6-}$:

$$\begin{pmatrix} \text{Si} & \text{O} \\ 2 \times +4 & 7 \times -2 = -6 \end{pmatrix}$$

A typical example is thortveitite ($Sc_2Si_2O_7$) which contains the scandium ion Sc^{3+}, but this is a rare mineral and this type is of little significance in weathering.

2.1.3 Two shared oxygen atoms

In contrast to the straightforward structures where no or one oxygen atom is shared when two oxygen atoms of the SiO_4 tetrahedra are shared, a number of possibilities arise.

Cyclic

These structures are called cyclic metasilicates and occur with three, four, six or eight joined SiO_4 tetrahedra. Three SiO_4 tetrahedra, each sharing two oxygen atoms, are shown in Fig. 2.3. The anion is $(SiO_3)_3^{6-}$:

$$(+4 - 6) \times 3 = -6$$

An example is bentonite, $BaTiSi_3O_9$, in which Ba^{2+} and Ti^{4+} provide the +6 charge needed to make the compound electrically neutral. In beryl, $Be_3Al_2Si_6O_{18}$, six SiO_4 tetrahedra each share two oxygen atoms in a ring structure. The charge on the $(SiO_3)_6^{12-}$ anion is counterbalanced by three Be^{2+} and two Al^{3+} ions.

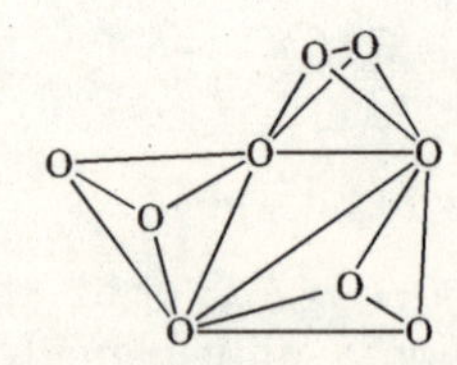

FIGURE 2.3 Cyclic metasilicate with three SiO_4 tetrahedra each sharing two oxygen atoms.

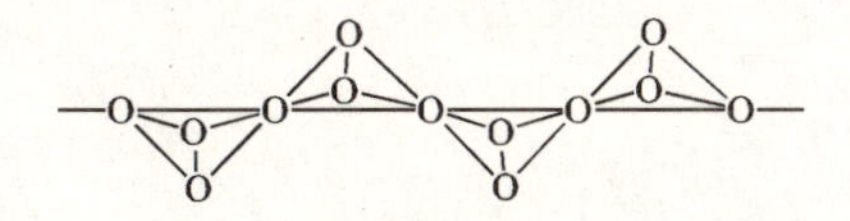

FIGURE 2.4 Part of a chain silicate with SiO_4 tetrahedra each sharing two oxygen atoms.

Both bentonite and beryl have properties typical of ionic solids. The sharing of three or six tetrahedra is more common than four or eight.

Chains

In this case the SiO_4 tetrahedra are joined together in very long chains (*see* Fig. 2.4). The generalised formula of these chain metasilicates is $(SiO_3^{2-})_n$ where n is a large number reflecting the length of the chain.

The *pyroxene* minerals (*see* Table 2.3) adopt this structure and an example is spodumene, $LiAl(SiO_3)_2$. Li^+ and Al^{3+} counterbalance the charge on $(SiO_3)_2^{4-}$. The chains are packed parallel to each other to give a fibrous-like structure and to create sites of either 6 or 8 coordination in which the cations are found.

The chain metasilicates exhibit wide structural variation because the tetrahedra may be linked to give repeat distances along the chain after between one and 12 tetrahedra.

Three possible variations, namely 1T, long 2T and short 2T, are shown in Fig. 2.5. The name '1T' means the repeat distance just involves one tetrahedron. 'Long 2T' means that the repeat distance involves two tetrahedra linked in such a way as to give a longer repeat distance of 520 pm as compared to 420 pm in the 'short 2T' arrangement (note that pm means picometres, which are 10^{-12} metres). The most common arrangement is the 2T structure.

Cross-linked chains

In these structures chains are cross-linked by shared oxygen atoms so that some tetrahedra share two and some share three oxygen atoms (*see* Fig. 2.6). The chains can have the same variation in the length of the repeat unit discussed in the previous section. The simplest is $Si_2O_5^{2-}$ based on the 1T structures found in sillimanite ($AlAlSiO_5$) which is an aluminium aluminosilicate. The most important group of minerals with cross-linked chains are

TABLE 2.3 Pyroxene minerals

Name	Formula
Diopside	$CaMgSi_2O_6$
Enstatite	$Mg_2Si_2O_6$
Jadeite	$NaAlSi_2O_6$
Spodumene	$LiAlSi_2O_6$

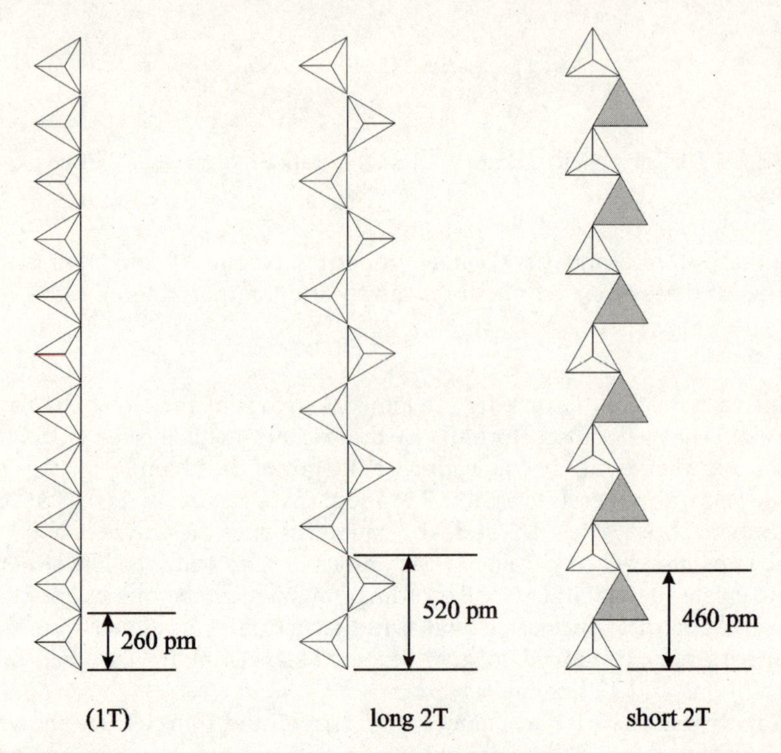

FIGURE 2.5 Three possible arrangements for chain metasilicates.

the *amphiboles* which are based on a 2T structure, $Si_4O_{11}^{6-}$, e.g. tremolite, $Ca_2Mg_5(Si_4O_{11})_2(OH)_2$.

These materials have hard fibrous structures held together by cations just as for pyroxenes. The best known of this group are asbestos minerals which are based on serpentines and amphiboles. White asbestos is the serpentine mineral chrysotile, $Mg_3(Si_2O_5)(OH)_4$.

2.1.4 Three shared oxygen atoms

This gives rise to sheet structures with the formula $(Si_2O_5^{2-})_n$ (*see* Fig. 2.7). The planar arrangement with the regular structure shown is rare, and

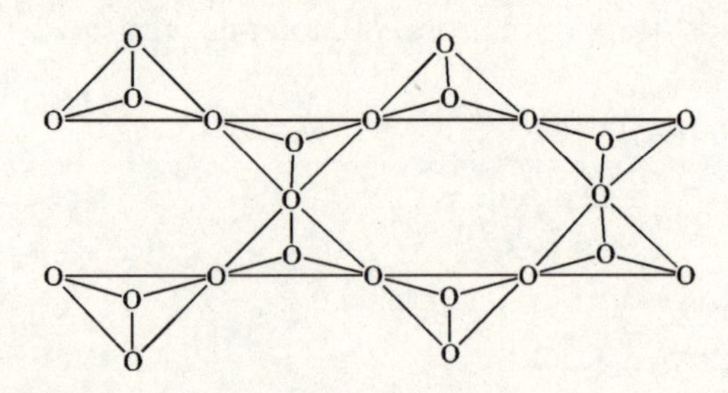

FIGURE 2.6 Cross-linked chain silicate.

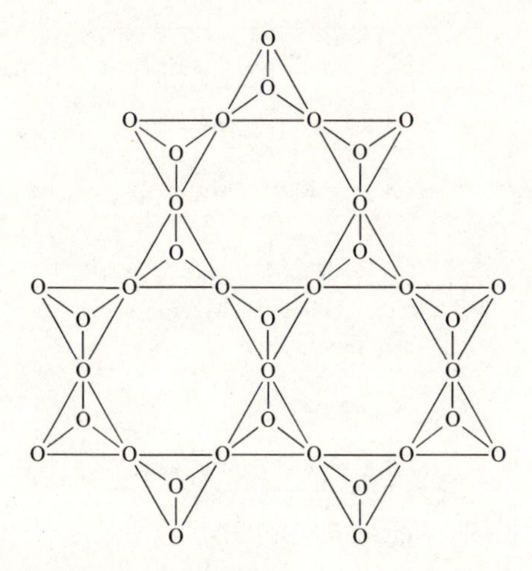

FIGURE 2.7 Sheet structure for $(Si_2O_5^{2-})_n$.

more usually puckered arrangements are found, or those based on alternate four-tetrahedra and eight-tetrahedra rings or with equal numbers of four-, six- and eight-tetrahedra rings.

The sheet structures encompass many important minerals. The *clay minerals* are of particular importance, e.g. kaolinite, montmorillonite and vermiculite; they are discussed in Chapter 10. Mica is a sheet structure aluminosilicate in which Al^{3+} replaces Si^{4+} at the centre of some of the SiO_4 tetrahedra to give layers which have the composition $AlSi_3O_{10}^{5-}$. The formula is $KMg_3(AlSi_3O_{10})(OH)_2$. In the schematic diagram (Fig. 2.8) the layers are clearly shown with the $AlSi_3O_{10}^{5-}$ layers held together by K^+ ions. Mica is hard but has pronounced cleavage planes which make for mechanical weakness and an ability for it to be split into sheets.

It is interesting to contrast mica with talc, $Mg_3(Si_2O_5)_2(OH)_2$. Talc is not itself an important mineral but certain features of its structure are significant in other key minerals (e.g. illite; *see* Chapter 10). A schematic diagram of talc is shown in Fig. 2.9. The forces between the layers are very weak as, unlike

Layer	
$AlSi_3O_{10}^{5-}$	
Mg^{2+} Mg^{2+} Mg^{2+} OH^- OH^- OH^- Mg^{2+} Mg^{2+} Mg^{2+}	Brucite layer
$AlSi_3O_{10}^{5-}$	Aluminosilicate layer
K^+ K^+ K^+	Potassium ions
$AlSi_3O_{10}^{5-}$	

FIGURE 2.8 Structure of mica, $KMg_3(AlSi_3O_{10})(OH)_2$.

$Si_2O_5^{2-}$ layer

Mg^{2+} Mg^{2+} Mg^{2+}
OH^- OH^- OH^- OH^-
Mg^{2+} Mg^{2+} Mg^{2+}

$Si_2O_5^{2-}$ layer

Weak forces holding the layers together
(therefore very soft mineral)

$Si_2O_5^{2-}$ layer

Mg^{2+} Mg^{2+} Mg^{2+}

FIGURE 2.9 Structure of talc, $Mg_3(Si_2O_5)_2(OH)_2$.

mica, there are no ions which hold the layers together. The nature of layer silicates is discussed more fully in Chapter 10, while Chapter 7 contains a review of hydration by water or aqueous solutions penetrating the layers and bringing about weathering.

Double layers

Double layers can be formed by adjacent silicate sheets sharing the fourth oxygen atom of the SiO_4 tetrahedra (*see* Fig. 2.10). This is found in structures where Al^{3+} replaces Si^{4+} in half of the SiO_4 tetrahedra, leading to aluminosilicates such as $CaAl_2Si_2O_8$ (*see* Chapter 10).

Another important double layer structure is the clay mineral kaolinite, ($Al_2(OH)_4Si_2O_5$), which arises from the weathering of feldspar ($KAlSi_3O_8$):

$$2KAlSi_3O_8 + 2H_2O + CO_2 \longrightarrow Al_2(OH)_4Si_2O_5 + K_2CO_3 + 4SiO_2$$

2.1.5 Four shared oxygen atoms

Framework structures occur for silica (SiO_2) and for aluminosilicates, the most important of which are the *feldspars*. The naming of feldspars is

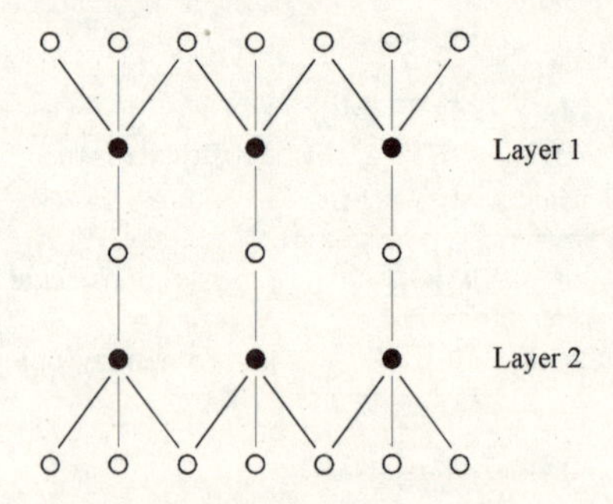

FIGURE 2.10 Sideways view of double layers.

Box 2.2 The naming of feldspars

The feldspars are members of a group of aluminosilicates, mainly of calcium, sodium and potassium. The term 'feldspars' is derived from the Swedish '*feltspat*' (1747). '*Felt*' means 'field' and refers to the importance of plant nutrients released during weathering. '*Spar*' is a miners' word for a well-cleaved (*see* Chapter 3), light-coloured mineral.

Feldspars are divided into two groups.

- *Plagioclase feldspars* consist of six members (Fig. 2.11) with varying amounts of sodium and calcium. The word 'plagioclase' was originally applied to feldspar which had a narrow angle between its two main cleavage planes. So the name in this case is based on crystal form.
- *Alkali feldspars* contain varying amounts of sodium and potassium. The word 'alkali' refers to the fact that sodium and potassium are both found in group I of the periodic table and that this group of elements is called the alkali metals because of the strongly basic nature of their oxides and hydroxides. In the case of the alkali feldspars the name is based on chemical composition. However, within this group the potassium feldspar, *orthoclase* ('straight splitting'), has a name based on crystal form as the cleavage planes intersect at 90°.

discussed in Box 2.2. In these structures, up to 50 per cent of the Si^{4+} ions have been replaced by Al^{3+}. The replacement of Si^{4+} by Al^{3+} means that cations need to be added to balance the negative change on the aluminosilicate framework. Series of feldspars exist in which one type of ion replaces another; for example, Na^+ in albite can be progressively replaced by K^+, ending up as orthoclase. These are the *alkali feldspars*:

$$\underset{\text{Albite}}{NaAlSi_3O_8} \longrightarrow Na_xK_{1-x}AlSi_3O_8 \longrightarrow \underset{\text{Orthoclase}}{KAlSi_3O_8}$$

Na^+ in albite can also be replaced by Ca^{2+}, ending up as anorthite. In this case Si^{4+} is replaced by Al^{3+}. These are the *plagioclase feldspars*:

$$\underset{\text{Albite}}{NaAlSi_3O_8} \longrightarrow Na_xCa_{1-x}Al_{2-x}Si_{2+x}O_8 \longrightarrow \underset{\text{Anorthite}}{CaAl_2Si_2O_8}$$

The variation in the make-up of the feldspars is shown in Fig. 2.11, which is a *ternary phase diagram* in which the composition of different feldspars is given. The construction of ternary phase diagrams is explained in Box 2.3. You should notice from Fig. 2.11 that the replacement of K^+ in orthoclase by Ca^{2+} does not occur and that mixtures of two feldspars are seen between these two aluminosilicates.

Silica (SiO_2), in which all of the four O atoms are shared (*see* Fig. 2.2) exists in more than 22 phases (different crystalline structures). The most common

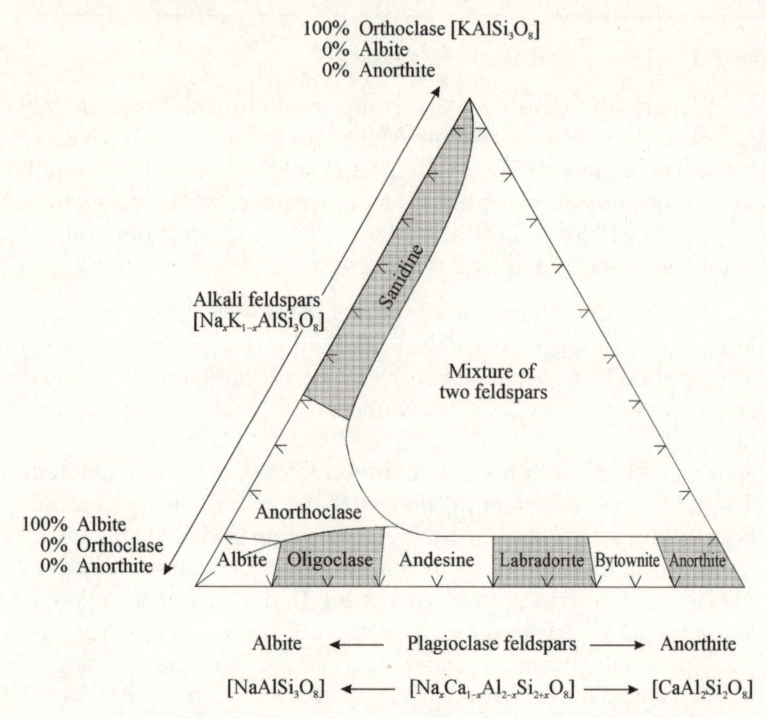

FIGURE 2.11 Ternary phase diagram for feldspars.

form is α-quartz, which is an important part of granite and of sandstone. You will probably know quartz best as beach sand, which is commonly brown due to the presence of iron oxide. The three-dimensional framework of the structure of SiO_2 held together by strong Si–O bonds ensures that quartz is very resistant to weathering processes. The structure of α-quartz consists of helical chains of tetrahedra of SiO_4 wrapped around each other.

2.2 CARBONATES

Calcium carbonate ($CaCO_3$) is of widespread occurrence in sedimentary deposits, which are often derived from the fossil remains of marine life. The mixed carbonate, dolomite [$MgCa(CO_3)_2$] is also widely distributed; for example, the Dolomites in Italy consist mainly of this mineral.

Calcium carbonate occurs in two forms, calcite and aragonite.

2.2.1 Calcite

This has a structure similar to that of sodium chloride (*see* Fig. 1.5) in which the planar carbonate ions (CO_3^{2-}) are in a cubic close-packed type of arrangement, while the calcium ions (Ca^{2+}) are in octahedral sites surrounded by six oxygen atoms provided by carbonate ions (*see* Fig. 2.13).

Box 2.3 Ternary phase diagrams

A ternary phase diagram is a two-dimensional figure used to show the composition of a system which contains three components. The sum of the three components must give a total composition of 100 per cent. Figure 2.12 is a simplified version of Fig. 2.11 and shows how a ternary diagram works.

Each corner of the equilateral triangle represents 100 per cent of a particular component, e.g. in Fig. 2.12 the top point of the triangle is 100 per cent orthoclase. The line opposite to that corner has 0 per cent orthoclase. To find the point on the diagram for 5 per cent orthoclase, 75 per cent anorthite and 20 per cent albite (remember the three must add up to 100 per cent), draw a line parallel to the 100 per cent albite - 100 per cent anorthite line at a value of 5 per cent orthoclase. You then draw a line parallel to the 100 per cent orthoclase - 100 per cent albite line at a value of 75 per cent anorthite. Where the two lines intersect gives the third value of 20 per cent albite. You can of course select any two of the three values to find the point on the diagram.

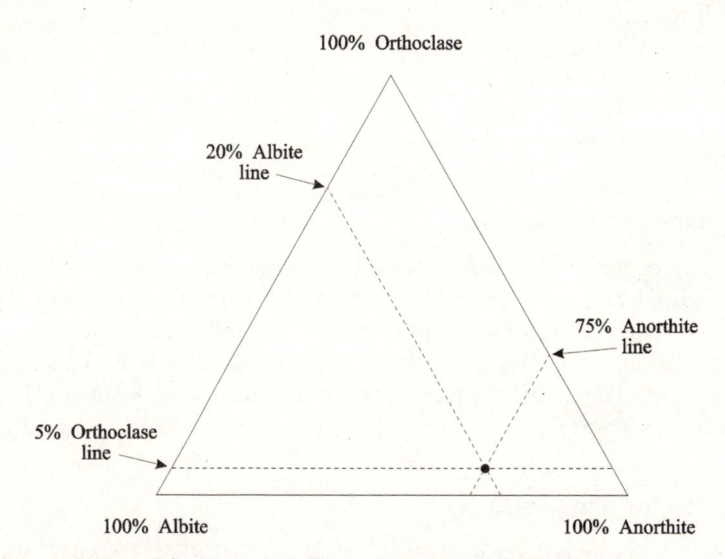

FIGURE 2.12 Simplified ternary phase diagram of feldspars.

2.2.2 Aragonite

In this structure the carbonate ions can be regarded as being in a hexagonal close-packed type of arrangement with the Ca^{2+} surrounded by nine O atoms provided by the carbonate ions.

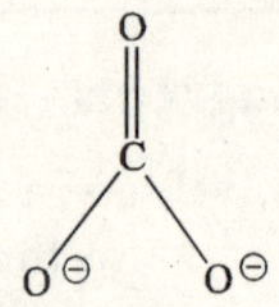

FIGURE 2.13 Planar carbonate ion (CO_3^{2-}).

Calcite is more common than aragonite, occurring as a major component of limestone (including chalk) and dolomite. Aragonite underlies extensive geographical areas such as the Red Sea basin and the Bahamas.

2.3 IRON OXIDE MINERALS

Iron is the fourth most abundant element in the Earth's crust (5.0 per cent; *see* Table 2.1). The main compounds of iron found in nature are the ores (particularly oxides) which are treated as products of weathering in Chapter 10. They are:

- hematite (Fe_2O_3)
- magnetite (Fe_3O_4)
- limonite ($2Fe_2O_3 \cdot 2H_2O$)
- siderite ($FeCO_3$)
- iron pyrites (FeS_2).

2.3.1 Hematite (Fe_2O_3)

Fe_2O_3 may occur in different crystalline forms (e.g. α- and γ- forms) but the important naturally occurring variety is hematite (α-Fe_2O_3). In this there is a hexagonal close-packed type of arrangement of oxide ions (O^{2-}) with iron(III) ions (Fe^{3+}) in two-thirds of the octahedral sites. This structure is called *corundum* which is named after α-alumina (α-Al_2O_3). γ-Fe_2O_3 is discussed in Chapter 10.

2.3.2 Magnetite (Fe_3O_4)

Magnetite contains both iron(II) (Fe^{2+}) and iron(III) (Fe^{3+}), unlike hematite which contains only iron(III) (Fe^{3+}). The formula of magnetite could be written as $(Fe^{2+})(Fe^{3+})_2O_4$. The oxide ions are arranged in a cubic close-packed type of arrangement with Fe^{2+} in octahedral sites and Fe^{3+} in both octahedral and tetrahedral sites. Magnetite is a part of a group of structurally similar minerals called *spinels*. The parent compound, spinel, is $MgAl_2O_4$. There are two types of spinel, each of which has a unit cell containing 32 O^{2-} ions with the metal ions M^{2+} and M^{3+} distributed as shown in Table 2.4.

Table 2.4 Normal and inverse spinels

Name	Number of ions in unit cell			Example
	Oxide (O^{2-})	M^{2+}	M^{3+}	
Normal spinel	32	8 in Td sites	16 in oct sites	Spinel, $MgAl_2O_4$
Inverse spinel	32	8 in oct sites	8 in Td sites 8 in oct sites	Magnetite, Fe_3O_4

Td = tetrahedral, oct = octahedral.

2.4 Compounds of aluminium

Aluminium forms a range of oxides and hydroxides which are discussed as products of weathering in Chapter 10. The oxide of aluminium is Al_2O_3 and in its naturally occurring form, which is α-Al_2O_3, it is called corundum (*see* section 2.3.1). The structures of the hydroxide and oxohydroxides are of interest because in addition to ionic bonding, hydrogen bonding also occurs (section 1.4).

2.4.1 Gibbsite (γ-$Al(OH)_3$)

There are double layers of hydroxide ions (OH^-) with Al^{3+} occupying two-thirds of the octahedral sites between the two layers. The double layers are then held together by hydrogen bonds. In addition to the hydrogen bonds between the double layers there are hydrogen bonds within each layer (*see* Fig. 2.14).

2.4.2 Diaspore (α-$AlO(OH)$)

The oxygen atoms in this structure are in a hexagonal close-packed type of arrangement with Al^{3+} in some of the octahedral sites. There are chains of edge-sharing AlO_6 octahedra which are built up in layers, and further connected by hydrogen bonds which are shown as double lines in Fig. 2.15.

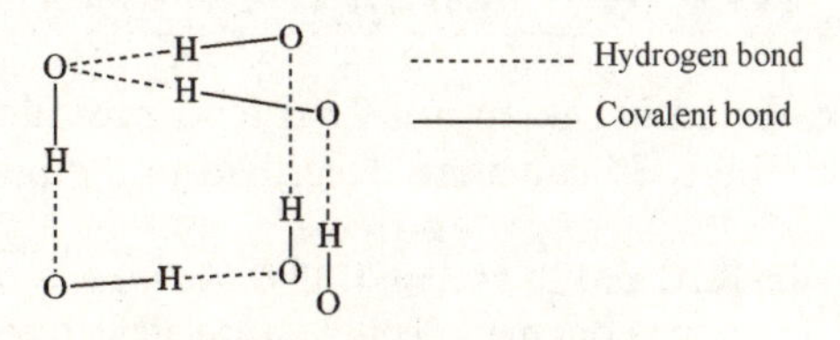

Figure 2.14 Hydrogen bonding in gibbsite ($Al(OH)_3$).

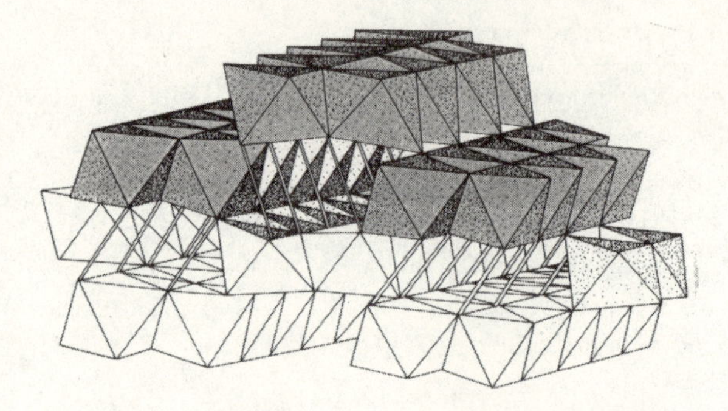

Figure 2.15 The structure of diaspore (α-AlO(OH)) showing AlO_6 octahedra in layers connected by hydrogen bonds. (Source: Greenwood, N.N. and Earnshaw, A. 1984: *Chemistry of the elements*. Oxford: Pergamon.)

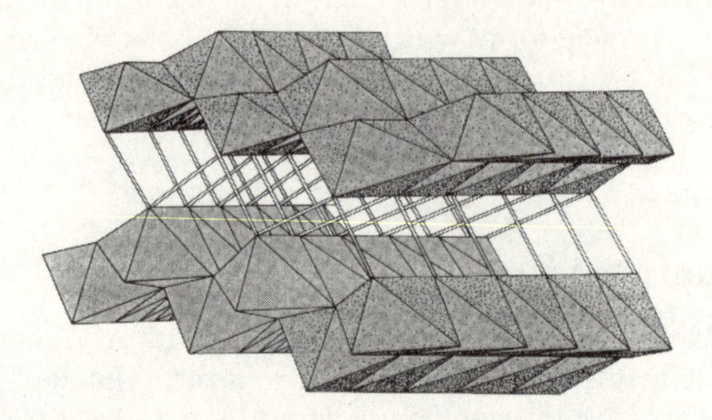

Figure 2.16 The structure of boehmite (γ-AlO(OH)) with hydrogen bonds shown as double lines. (Source: Greenwood, N.N. and Earnshaw, A. 1984: *Chemistry of the elements*. Oxford: Pergamon.)

2.4.3 Boehmite (γ-AlO(OH))

This has a layer structure. Within each layer the O^{2-} ions are in a cubic close-packed type of arrangement, though overall the O^{2-} ions are not in a close-packed type of structure. Hydrogen bonds, shown as double lines in Fig. 2.16, hold the structure together.

Readings for further study

The two textbooks listed at the end of Chapter 1 provide some support for the work in this chapter. Silicates are described in the books given below.

Deer, W.A., Howie, R.A. and Zussman, J. 1992: *An introduction to the rock forming minerals*, 2nd edn. London: Longman. This is a detailed reference book of rock-forming minerals.

Greenwood, N.N. and Earnshaw, A. 1997: *Chemistry of the elements,* 2nd edn. Oxford: Butterworth Heinemann. This textbook contains a clear and comprehensive description of the structure of silicate minerals.

Wells, A.F. 1984: *Structural inorganic chemistry,* 5th edn. Oxford: Oxford University Press. This large reference book (over 1000 pages) is an excellent source book which provides a wealth of structural data for solid compounds.

3

BULK PHYSICAL PROPERTIES OF ROCKS

Rocks at and near the Earth's land surface show a variety of responses to weathering processes. At a simple level this variety is due to the range of climatic environments and to differences in the nature of rocks. Rock differences reflect

- *chemical and mineralogical properties* which influence chemical weathering processes (discussed in Chapters 1 and 2);
- *bulk physical properties* which may encourage or resist both physical and chemical mechanisms (reviewed in this chapter).

The dividing line between these two sets of properties is unclear. We have drawn it somewhere above the molecular scale, and so have used the term 'bulk' to describe physical properties.

The first point to note about the physical properties of a rock is that they tend to vary spatially. For example, the presence of discontinuities (see below) will mean abrupt changes in the strength of a rock as measured along a line through the rock which crosses such discontinuities. A rock which shows variation in physical properties along a direction is said to be *anisotropic* (*see* Box 3.1).

Box 3.1 Isotropy and anisotropy

Isotropy is a condition of a substance, such as a synthetic alloy (e.g. steel) or a ceramic, when it shows the same properties in all directions. *Anisotropy* is the condition of a substance when it shows different characteristics depending on a direction. Most rocks are anisotropic as far as their physical characteristics are concerned. The compressive strength (see below) of a schist, for example, may be 70 per cent weaker along its layers than normal to them. (A schist is a layered metamorphic rock, typically rich in mica.)

Table 3.1 Physical rock properties relevant to weathering

Inherent	Mechanical
1. Texture: including grain size and its relation to porosity, surface area, permeability and capillarity	1. Compressive strength
2. Discontinuities at various scales	2. Shear strength
3. Water content, measured by natural moisture content; water absorption capacity; saturation coefficient	3. Tensile strength
	4. Elasticity

The physical properties that are relevant to weathering studies may be divided into 'inherent' and 'mechanical' types (*see* Table 3.1). *Inherent* properties are those related to the constituents and structure of a material, while *mechanical* properties describe the way a rock responds to changes in stress. The latter are important as they provide an index to the vulnerability or otherwise of a rock to the stresses set up during physical weathering especially.

3.1 Texture

Texture is the relationship between the mineral grains that form a rock. The nature of this relationship is affected by the size and shape of the grains and by the nature of the contacts between them. Texture influences a rock's response to weathering through its effect on rock strength, and so resistance to physical weathering, and more particularly through its control of water availability and movement.

Two types of texture are significant for weathering, crystalline texture and sedimentary texture.

3.1.1 Crystalline texture

In this case the constituent particles of a rock consist of interlocking crystals. This is typically the case for many unweathered igneous rocks, whose component minerals crystallised at different temperatures as the original molten fluid ('magma') cooled and consolidated. A rock with a crystalline texture is generally resistant to weathering. It has a high resistance to stress, shows a large range of elastic behaviour (see below) and has a low porosity (*see* Fig. 3.1), perhaps 0.1 per cent in the case of granite.

A variety is *poikilitic texture*, which occurs when a large crystal completely encloses one or more that are much smaller. The smaller crystals may consist of easily weathered minerals, but will remain unaltered if the surrounding mineral is resistant. The parent rock will therefore resist chemical weathering even though it may contain a high proportion of readily altered minerals.

3.1.2 Sedimentary texture

This is particularly important as sedimentary rocks underlie about 80 per cent of the continental surfaces. It contrasts with the crystalline variety in that the component particles do not generally show interlocking characteristics. You will find that a study of the implications of *grain size* provides a helpful key to an understanding of both the content of water and its movement along discontinuities (*see* section 3.1.2) in sedimentary rocks.

Relationships between grain size and the properties of sorting, porosity, surface area and water content

Grain size affects the amount and potential activity of water through its relationship with sorting, porosity and surface area. You will find it quite easy to follow the ideas if we start with the simple situation of an uncemented sediment where all the grains are rounded and are about the same size, i.e. they are all *well sorted* (*see* Fig. 3.1). If the grains are packed together (section 1.1.4) in the most efficient manner, i.e. they occupy the smallest volume, then the amount of empty space may be as low as 26 per cent of the total. This proportion is the *porosity*. If, however, they are packed together in the most inefficient way, i.e. they occupy the maximum volume, then the porosity may be as high as 48 per cent.

These simple ideas now need some expansion and modification.

You should note first that while a reduction in grain size has no effect on porosity, it does increase the available surface area within a unit volume of material. This idea is explained further and illustrated in Box 3.2. The second point to note is that, in nature, particles are rarely perfectly sorted. Smaller grains typically occupy empty spaces among larger particles, so the overall sorting may be poor and the porosity low. There is a tendency for finer sediments to be better sorted than coarser ones, and so to have a higher porosity. A silt (0.004–0.0625 mm diameter) may be better sorted, and have a higher porosity, than a gravel (2–60 mm diameter).

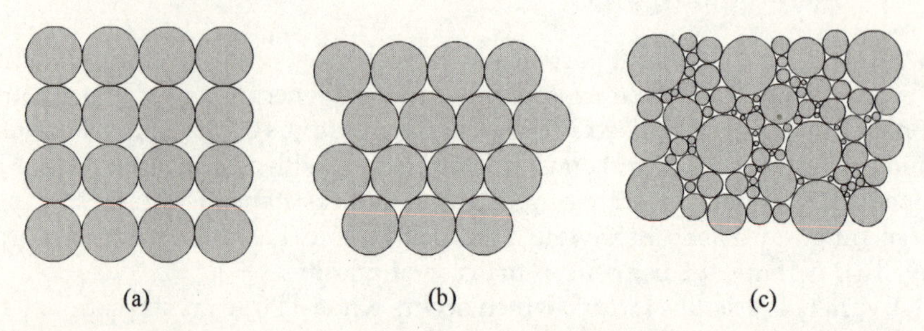

FIGURE 3.1 Sorting, packing and porosity. (a) Well sorted but poorly packed. Relatively high porosity (maximum possible in three dimensions is nearly 48 per cent). (b) Well sorted and well packed. Relatively low porosity (minimum possible in three dimensions is nearly 26 per cent). (c) Poorly sorted with low porosity. Note that porosity is measured as (volume of voids)/(total volume).

Box 3.2 The relationship between particle size and surface area for a fixed volume of material

As the constituent particles of a material *decrease* in size, the total surface area of the particles, i.e. the zone of potential water–mineral interaction, *increases*.

1. A cube of 1 cm sides has a surface area of **6 cm^2**.
2. Now consider a similar cube made of silt-sized particles (spheres) each of radius 10^{-3}cm (0.02 mm diameter), and showing best packing (see section 1.1.4), i.e. 26 per cent porosity.
   It follows that 74 per cent (i.e. 0.74 cm^3) of available space is occupied. **One** silt particle has a volume of $\frac{4}{3}\pi r^3$ cm^3. So
   $$n(= \textbf{total number of particles}) \times \tfrac{4}{3}\pi(10^{-3})^3 = 0.74$$
   i.e.
   $$n = \frac{3 \times 0.74}{4\pi \times 10^{-9}}$$
   Now the surface area of *each* particle $= 4\pi r^2$, $4 \times \pi \times (10^{-3})^2$ cm^2. So
   $$\text{the } \textbf{total surface area} = n \times \text{the individual area}$$
   $$= \frac{3 \times 0.74 \times 4\pi \times 10^{-6}}{4\pi \times 10^{-9}} = \mathbf{2220\ cm^2}$$
3. The thin (7–14 Å; note that 1 Å = 10^{-8} cm) and platey nature of clay mineral crystals implies a high surface area; 1 cm^3 of dry montmorillonite has a surface area of about **300 m^2**.

The two properties of porosity and surface area have major significance for weathering. A high porosity suggests a potentially large water content with implications for, for example, frost-based weathering. A large surface area implies an extensive interface between water and grain and so enhanced chemical weathering. In nature, water availability will influence the percentage of pore space that is filled, and the extent of water–grain contact. This factor is dealt with below. In addition, two internal factors influence water content.

Degree of void linkage

Unless water can travel between voids, a figure for rock porosity may not be particularly helpful as a measure of likely water content.

Presence of cementing minerals

The 'theoretical' space between grains is often occupied by a cementing material, perhaps precipitated by earlier circulating water. Three types of cementing material are commonly found:

Box 3.3 Porosity variations in the Carboniferous Limestone of Northern England

The main limestone types show a range of porosities.

- *Micrites* have a very low porosity of 2 per cent or less. They have a very dense groundmass (cement) made up of very small calcite particles and clay-paste carbonate.
- *Sparry limestones* (*sparites*) have a porosity of about 5–8 per cent. The high content of crystalline calcite is associated with reduced pore space.
- *Biomicrites* have the highest porosity of 15–25 per cent. The groundmass is mainly made of granular and fragmented matter, allowing for many voids.

As a comparison chalk has a high porosity, nearly 50 per cent.

(Based on: Sweeting, M.M. 1970: Recent developments and techniques in the study of karst land forms in the British Isles. *Geographia Polonica* **18**, 227–41.)

- *calcium carbonate,* which may bind the components of a limestone, or superficial sands and gravels over limestone;
- *iron oxide minerals;*
- *silica.*

While porosity may be lowered by the presence of a cement, it may be increased through the dissolving action of percolating water. The range of porosities in a single rock group may therefore be quite large and this is illustrated in Box 3.3.

Relationship between grain size and permeability

While pore space per unit volume tends to increase as particle size decreases, the *permeability* of a rock, i.e. its ability to transmit a fluid (normally water), tends to decrease. For example, a clay may transmit water at a rate of about 10^{-6} m day^{-1}, while a coarse gravel may have a permeability of 10^3 m day^{-1}. When a sediment shows a regular variation in particle size characteristics, as would be the case with graded bedding (*see* Fig. 3.2), then local zones of enhanced water flow would subsequently occur. This pattern clearly has implications for the response of such a sediment to water-based weathering processes.

Relationship between grain size and capillarity

You may have noticed how a relatively fine and porous material like dry sand becomes damp when part of it is brought into contact with water.

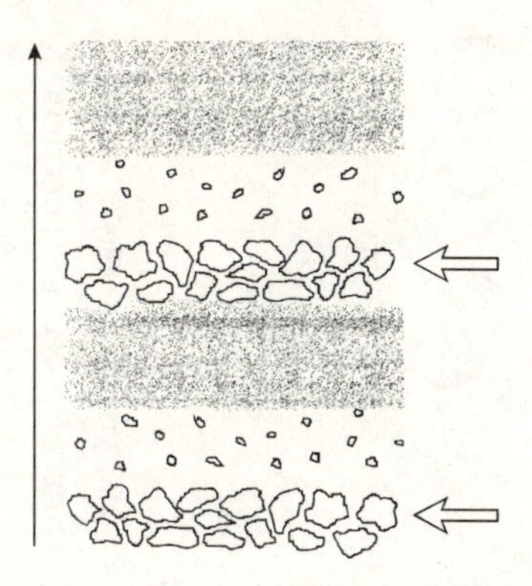

FIGURE 3.2 Graded bedding. The vertical arrow indicates the direction of sediment accumulation during periods of fluctuating water-flow energy. Coarser debris is deposited at high flow velocities, and finer as flow lessens. The horizontal arrows identify coarse bands of sediment having relatively high permeability, and so much scope for water-based activity.

This phenomenon is an example of *capillarity*, which is the tendency of most liquids (notably water, the important liquid at and near the land surface) to rise up a narrow tube placed in open water.

In order to understand the process you should become familiar with the forces acting within water, and between water and adjacent material. The attractive forces within water, which give rise to *cohesion*, are stronger at the surface than at depth. This is because the hydrogen bonds offered by surface molecules are largely directed inwards, away from the few water-vapour molecules in air. This asymmetry does not apply to molecules in the interior of the water. The high degree of cohesion between surface molecules and those below gives rise to *surface tension*. The second force, *adhesion*, acts between the molecules of water and those of the containing material.

These forces bring about the capillary rise of water in tubes. Adhesive forces, aided by surface tension and opposed by gravity, drag water up the tube walls. The greatest rise is in tubes of very small diameter because, in this situation, adhesive forces are most dominant over the restraining gravitational forces. A consequence is that the water surface, or *meniscus*, is sharply curved, i.e. it has a very low *radius of curvature*. The height to which water rises is given by

$$h = \frac{2\gamma}{gr}$$

where γ is the surface tension, g is the gravitational acceleration and r is the radius of curvature of the meniscus.

In the case of a regolith, linked pores take the place of tubes (Fig. 3.3). The average particle size determines the diameter of the tubes and thus the curve of the meniscus. Size is quite easily measured, and so it is a useful surrogate

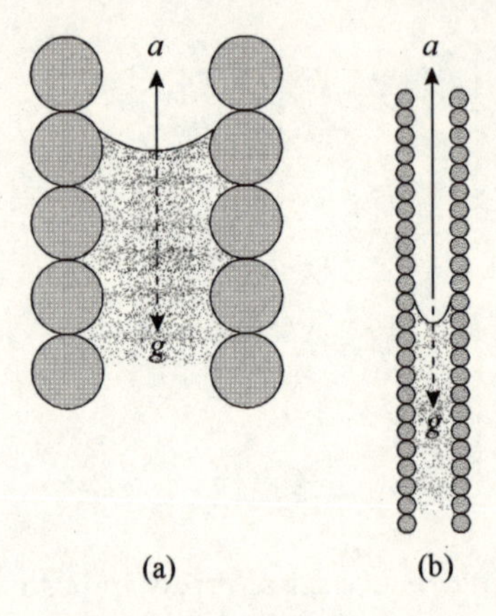

FIGURE 3.3 Capillarity in unconsolidated material. (a) 'Tubes' in coarse particles have a relatively large diameter. The ratio between surface area of water and its contact rim is high, and the modest upward force *a* hardly offsets the gravitational force *g*. (b) 'Tubes' in silt-sized particles are relatively narrow. The ratio between the surface area of water and its contact rim is low, and the high upward force *a* overwhelms the gravitational force *g*.

(replacement measure) for radius of curvature. Experiments on various materials show that water may be lifted through coarse sand to a height of about 0.5 m above the open-water level. An elevation of about 50 m is possible in fine silt (0.01–0.001 mm diameter), but this would take a very long time.

Very similar principles apply when very narrow cracks ('microcracks') occur instead of 'tubes'. Water migration again takes place; a rate of 25 mm h^{-1} has been recorded in microcracks in basalt. The migration of water may have significant consequences for weathering.

3.2 DISCONTINUITIES

The anisotropic nature of many rocks is expressed through the frequent presence of mechanical discontinuities (see Box 3.4 for an example of how they may be described). Discontinuities are breaks or fractures of varying scale and may be open, closed or sealed. In the latter two cases a plane of mechanical weakness may still exist. Discontinuities modify the response of a rock or mineral to weathering in three main ways.

- Within a given volume of material they increase the area that is likely to be in contact with water. A consequence is that the rate at which a unit of materials responds to chemical weathering is likely to be increased.
- They weaken the resistance of rock in the mass to mechanical stress which may for example be exerted through the freezing of water in fractures.

Box 3.4 Rock quality designation (RQD)

An RQD value is used in civil engineering to describe the spacing of discontinuities in a rock. The RQD is the ratio between total unbroken lengths (TUL) usually over 100 mm (RQD_{100}) and the overall length usually 1 m, expressed as a percentage.

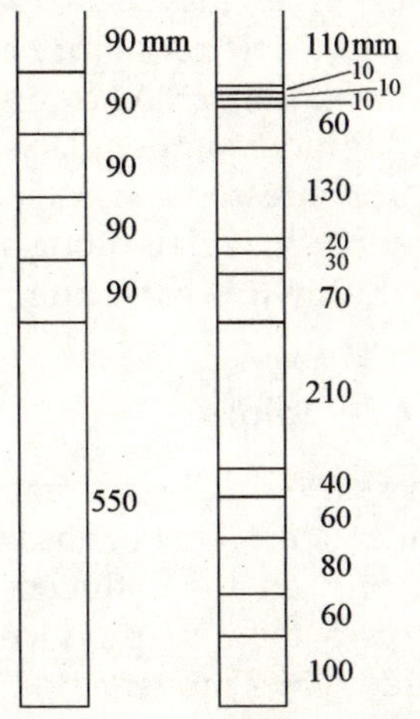

The two profiles show how the selection of TUL distance (here 100 and 300 mm) can affect the value.

		110
		+130
		+210
TUL for		+100
RQD_{100}	550	550
RQD_{100}	55	55
TUL for RQD_{300}	550	0
RQD_{300}	55	0

(Based on: Hawkins, A.B. 1986: *Site investigation practice: assessing BS5930.* Engineering Geology Special Publication, London: Geological Society.)

- At the smaller (microcrack) scale they may contain weathered minerals that expand on wetting and so induce further failure. In addition, contained water may freeze and also bring about breakdown.

Discontinuities are found at a wide range of scales from 10^3 to 10^{-9}m (kilometre to nanometre). It is convenient to discuss their significance for weathering in approximate order of size.

3.2.1 Faults

These are fracture surfaces in a body of rock along which displacement of adjacent blocks has taken place. Faults modify the response to weathering of local rocks in two main ways:

- They are lines (or zones, where faulting is intense) of high permeability and so they increase the area of rock exposed to water. The movement of water may be enhanced when faulting brings permeable and less permeable rocks into contact. Permeability is also aided by the frequent development of complex fractures associated with the main fault. In addition, joint density often increases in the vicinity of major faults.
- The stresses generated by faulting may lead to the production of new, softer materials along a fault plane or zone. These materials may respond readily to weathering.

3.2.2 Joints

These are small-scale fractures that result when a rock shows brittle failure under a tensional or shearing stress (see below). The type of stress has some bearing on the influence of a joint on weathering processes. Tensional stresses tend to produce joints that are open and so provide zones of ready water movement. On the other hand, shearing stresses tend to result in joints that are tightly closed and so much less permeable. The openness of a joint may be called its 'thickness'.

As well as degree of openness, two other geometrical characteristics affect the response of jointed rock. The first is joint spacing, usually expressed as the average distance between joints. The closer the spacing between joints, the more rapidly will a rock mass be reduced by weathering, simply because of the large area of potential water-contact surface per unit volume of rock. The second characteristic is joint continuity. Where this is well developed, permeability is enhanced, mechanical strength is reduced and susceptibility to weathering is increased.

The spatial pattern of joint systems has a marked influence on the surface forms of a rock such as granite (*see* Chapter 11).

3.2.3 Bedding planes

These are surfaces that separate beds in sedimentary rocks. They represent breaks in deposition and so are discontinuities, showing a tendency to be open towards the ground surface where the compressive stress (see below) exerted by the overburden is low, and to be closed at depth. They are zones of variable water flow, which is often associated with the precipitation of iron and manganese oxides. Their mechanical weakness means that they are readily exploited by physical weathering processes.

3.2.4 Rock cleavage

When a rock is subject to a high deforming stress it may develop a planar fabric called rock cleavage. This consists of a set of fractures along closely spaced parallel surfaces, and is due to a preferred orientation of platy minerals such as mica, muscovite and chlorite. Degrees of cleavage are recognised which depend upon the mineralogy of the deforming rocks and on the magnitude of the applied stress.

Cleavage is a significant property influencing weathering as it allows water to penetrate along the fracture surfaces, which are also planes of mechanical weakness.

The importance of cleavage is well illustrated by Snowdonian slate. This is impermeable at right angles to the cleavage planes, but water readily penetrates many centimetres along these surfaces to cause alteration shown by the reddish staining of oxidation.

3.2.5 Mineral cleavage

Many minerals are mechanically anisotropic in that they have a tendency to split relatively easily and uniformly along one or more preferred directions. The smooth and plane surfaces that result are called *cleavage planes*. They coincide with planes within the atomic structure where bond strengths are at their weakest. For example, mica cleaves readily along parallel planes where large cations such as potassium weakly bond adjacent sheets of silicate tetrahedra (*see* Chapters 3 and 10). This mechanical anisotropy, which may be exploited by the stresses of weathering, varies between minerals.

Quartz has no cleavage planes, and so is mechanically strong. Mica has a single perfect cleavage and can be split easily into very thin sheets. Olivine has two poor crystal planes. Feldspars show two cleavages, as do pyroxenes and amphiboles. Grains of the mineral hornblende, a member of the amphibole group, may show 'cockscomb' terminations, partly due to selective solution along cleavage planes. Calcite has three cleavage planes and as a consequence tends to yield lozenge- or diamond-shaped cleavage fragments.

3.2.6 Microcracks

Small-scale physical discontinuities may occur within and between the mineral grains that make up many rocks. Such microcracks fall into three categories:

- *multigrain cracks* which cross several crystals, usually along cleavage planes or crystal boundaries;
- *single-grain cracks* which only involve single minerals and appear mainly as cracks along cleavages or along grain boundaries;
- *stress fractures* which are less influenced by mineralogy, and which are generally larger than the preceding two: they appear to be caused by rock stresses, and may be filled with secondary crystals.

Olivine and feldspar minerals are particularly associated with microcracks. Olivine, which has no cleavage, often shows small basal fractures and prismatic partings (*see* Fig. 3.4). Some microcracks may have been produced by stresses caused by rapid cooling as the mineral formed.

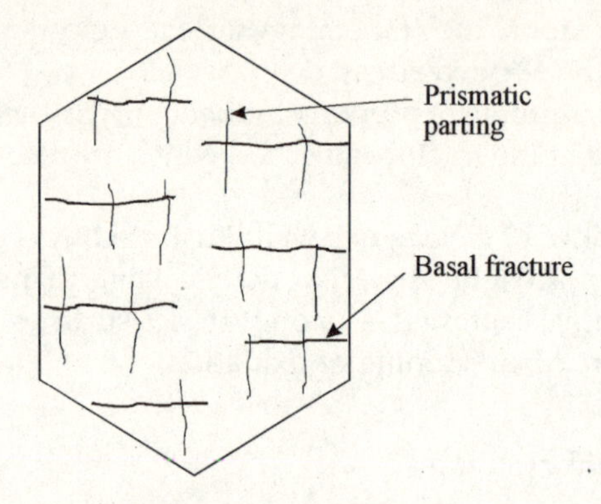

FIGURE 3.4 Microfractures in an olivine crystal.

Cracks may be exploited and extended as the olivine alters to the higher-volume serpentine during weathering. Generally, the disruptive effects of salts (*see* Chapter 6) and of swelling secondary minerals (*see* Chapter 10) can enlarge and extend microcracks.

3.3 MINERAL SURFACE COMPLEXITY AT THE VERY SMALL SCALE

In this section we are dealing with phenomena which are transitional between those at the bulk and at the atomic scales discussed in Chapters 1 and 2. Box 3.5 gives a brief outline of some of the units of measurement used at this scale.

The application of new techniques (*see* Box 3.6 for an outline of some of the tools used) is revealing the complex nature of mineral surfaces at the smallest scale. This complexity, expressed through composition, atomic structure and surface relief, is most important because it provides evidence of the conditions under which chemical weathering actually begins.

Box 3.5 Units for measuring very small lengths

Lengths are generally measured in multiples or fractions of a metre (m). For chemical weathering studies, where processes and forms may need to be investigated at the atomic or molecular scale, three units are often used:

- the *nanometre* (nm): $1\,\text{nm} = 10^{-9}\,\text{m}$
- the *angstrom* (Å): $1\,\text{Å} = 10^{-10}\,\text{m}$ (or 10^{-8} cm in older textbooks)
- the *picometre* (pm): $1\,\text{pm} = 10^{-12}\,\text{m}$

Box 3.6 Some modern tools for determining the composition, very small-scale form (microtopography), and atomic structure of mineral surfaces

Composition

- *X-ray photoelectron spectroscopy (XPS)* is the most widely used tool for surface analysis. It allows the identification of an element, its oxidation state and, in some cases, structural details.
- *Auger electron spectroscopy (AES)* is almost as important as XPS in surface analysis. It has been used to distinguish between true adsorption and surface precipitation.
- *Secondary ion mass spectrometry (SIMS)* involves the bombardment of a surface with ions, allowing the analysis of all elements to trace level concentrations (parts per billion).

Microtopography

- *Scanning electron microscopy (SEM)* is perhaps the most useful technique for the study of surface form. It can resolve features down to 10–20 Å.
- *Scanning tunnelling microscopy (STM)* can be used to scan areas as small as several square nanometres. Its use has provided information about the positions of microtopographic features.

Atomic structure

- *Low-energy electron diffraction (LEED)* is a technique for determining the atomic structure of surfaces.
- *Electron tunnelling spectroscopy (ETS)*, in combination with STM, provides information about surface electronic structure.

The *composition* of a mineral surface is not the same as that of the bulk of the mineral, and also varies across the surface. The difference arises partly because elements from the air or adjacent solutions (e.g. hydrogen, oxygen, carbon and sometimes nitrogen) attach themselves to an initially highly-reactive surface. A second reason for variation in composition is that chemical weathering may form a leached layer, due to a loss of ions, between a few and several thousand angstroms thick. These compositional changes affect the reactivity of the surface.

The *atomic structure* of the surface of a mineral is frequently not representative of that of the bulk. Typically, surface atoms shift inwards towards the bulk to achieve a lower-energy arrangement. In the case of a feldspar the spacing between the first and second atomic layers may be reduced by up to 15 per cent. There are implications for reactions with external ions. The

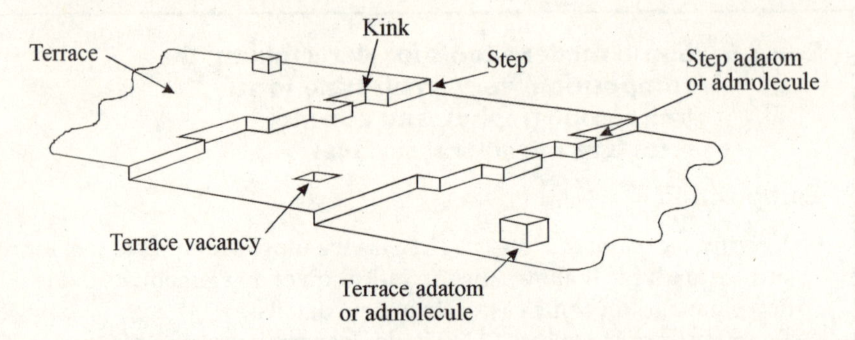

FIGURE 3.5 A diagram showing the main features of mineral surface microtopography. Each block represents a single atom, a group of atoms or a molecule. (Based on: Hochella, M.F. and White, A.F. (eds) 1990: Mineral–water interface geochemistry. *Reviews in Mineralogy* **23**.)

third important surface characteristic, and which also affects weathering response, is the presence of very small-scale relief, or *microtopography*. Surfaces which appear to have a mirror-like flatness when viewed through the petrographic microscope, such as those provided by the perfect cleavage of the lead sulphide galena (PbS), turn out to have considerable microtopography.

For example, an albite (a plagioclase feldspar having the formula $NaAlSi_3O_8$) cleavage surface may have vertical relief between 5 and 300 Å, and flat (at this scale) areas over 1000 Å wide.

A general model (Fig. 3.5) of mineral surface microtopography has been developed. This shows a number of relief components. 'Flat' areas are called *terraces*. They are separated by *steps*, which may be one or more atomic layers high. A *kink site* is found where a step changes direction. A *terrace vacancy* is a hole in a terrace the size of an atom or molecule. An atom or molecule sitting on top of a terrace is called an *adatom* or *admolecule*. A single atom extending from a step is a *step-adatom*. The most reactive surface localities are where the atoms have fewest neighbours (i.e. are adatoms), and the least reactive sites are where the atoms have most neighbours (i.e. terrace atoms). Surface geometry, then, will affect chemical weathering processes.

3.4 WATER CONTENT

The moisture content of a rock is determined by water availability as well as by internal space. Rocks are generally saturated below the water table (the *phreatic zone*), but above (the *vadose zone*), and in humid temperate regions, they are almost always less than 50 per cent saturated. There are four useful measures of water content. Box 3.7 gives details on how some of them are calculated.

Box 3.7 The determination of two measures of water content

1. **Water absorption capacity, *WA*:**

$$WA = \frac{W_3 - W_0}{W_2 - W_1}\%$$

where W_0 is the dry weight of sample after tests, W_1 is the weight of saturated sample suspended in water, W_2 is the weight of saturated sample in air and W_3 is the weight of sample that is first dried, then soaked for 24 h, wiped, and weighed in air.

2. **Saturation coefficient, *S*:**

$$S = \frac{W_3 - W_0}{W_2 - W_1} \times \frac{W_2 - W_1}{W_2 - W_0} = \frac{W_3 - W_0}{W_2 - W_0}$$

3.4.1 Natural (or field) moisture content

This is the water content of rock samples as they are collected in the field. It is the ratio of the weight of water, including surface water (often a substantial fraction of the total) to the weight of the dried sample. *Saturation moisture content* is the same ratio for a sample fully saturated under a vacuum in the laboratory. *Water absorption capacity* is a measure of the amount of water absorbed in a given time. *Saturation coefficient* is the amount of water absorbed in 24 h expressed as a fraction of the volume of available pore space.

It is important that the design of experimental studies of water-based weathering takes into account the actual field conditions of water availability.

3.5 Mechanical properties

These are the various responses of rocks to the different kinds of stress (*see* Box 3.8) that may act on them. Some stresses are due to the Earth's internal processes such as earth movements or, at a smaller scale, the cooling of formerly molten rock. Other stresses are set up by weathering processes such as temperature change or the conversion of pore water to ice.

In any event, a major consequence of stress for weathering is the growth of cracks at various scales. Such cracks may be exploited by weathering agents. The types of stress (*see* Box 3.8) and rock responses are summarised below.

Box 3.8 Stress

Definition and units

Stress is force per unit area. It is normally measured in newtons (N) per square metre. A stress of 1 newton per square metre ($1\,N\,m^{-2}$) is called a pascal. For strong materials such as rocks, stress may be measured in MPa (10^6 Pa) or MN (10^6 N) m^{-2}.

Types of stress

(a) Compressive, (b) tensile and (c) shear (arrows show direction).

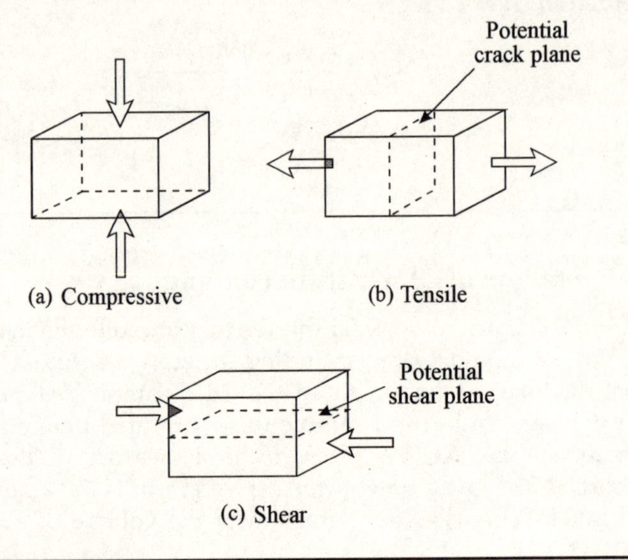

3.5.1 Types of stress

Compressive stress may be set up when a rock contracts during cooling, while a *tensile stress* may be applied when a rock expands in heating, or when the confining pressure of an overburden is released. *Shearing stresses* may be set up when an uneven force, perhaps due to temperature variation, is applied.

3.5.2 Rock response

Rocks respond by showing deformation, or *strain*, when a sufficient stress is applied. Four types of strain are relevant in weathering (*see* Fig. 3.6).

Elastic strain

In this case any strain due to stress is recoverable once the stress is removed. The elasticity of a rock is measured by a property called the *modulus of*

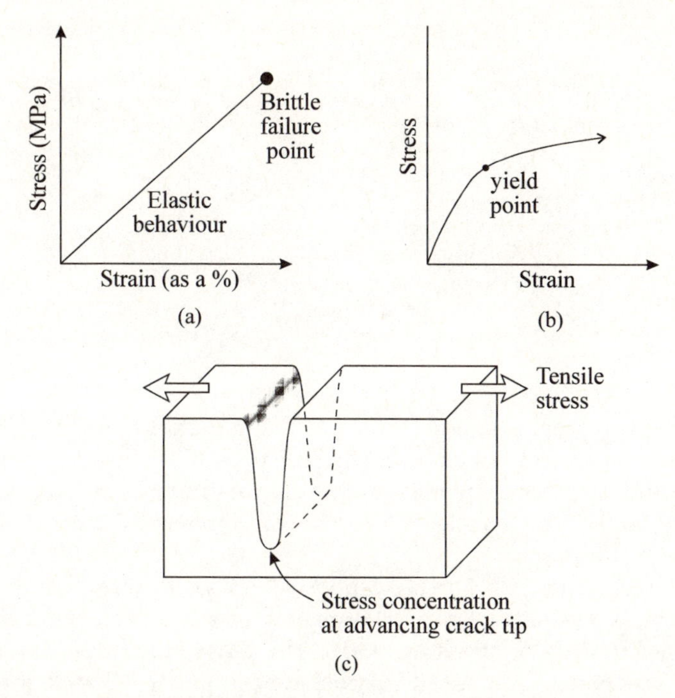

FIGURE 3.6 Types of strain. (a) Elastic deformation to brittle failure, (b) plastic failure and (c) crack development as tensile stress leads to brittle failure. Note: as rocks are generally strong, stress is often measured in megapascals (MPa; 1 MPa = 10^6 pascals).

elasticity, or *Young's modulus* (E). This expresses the relationship between stress (r) and strain (e), so that

$$E = \frac{r}{e}$$

It is recorded in units of stress ($N\,m^{-2}$) and is usually constant to the failure point. You can see that the greater the value of E, the less is the deformation produced by a given stress. Young's modulus is a measure of a rock's ability to resist deformation. Certain rocks may behave elastically when exposed to temperature change and so do not break down.

Brittle failure

If stress builds beyond the *elastic limit* (the point where strain ceases to be recoverable), a rock may show sudden catastrophic failure. Experimental results (Table 3.2) show that the level of failure (i.e. the yield strength) varies considerably within a rock type, between rock types and with the type of stress.

Failure in response to tensile stress is particularly important in weathering. The data in the table demonstrate that rocks generally show least

Table 3.2 Strength ($MN\,m^{-2}$) at failure for different rock types and stresses

	Compressive stress	Tensile stress	Shear stress
Granite	100–250	7–25	14–50
Dolerite	100–350	15–35	25–60
Basalt	100–300	10–30	20–60
Quartzite	150–300	10–30	20–60
Sandstone	20–170	4–25	8–40
Shale	5–100	2–10	3–30
Limestone	30–250	5–25	10–50

resistance to a tensile stress, although the variations are considerable. Resistance may in fact fall close to zero if a rock mass has discontinuities, but it should also be remembered that in weathering environments the expansion implied by tensile failure may be limited by the constraining effect of surrounding rock.

Crack development is a typical response when yield strength is exceeded. Such a crack grows normally to the direction of stress, and extends because of the concentration of stress at its tip. The tensile strength of a body of rock may be severely weakened by crack development. A crack about one-hundredth of a body's dimensions can reduce its tensile strength by 10^2–10^3 times.

Fatigue failure

This type of failure may occur when a stress is repeatedly applied, even though the magnitude of each stress event is well below the yield strength of the material. The strength after repeated stresses is the *fatigue limit*, which is usually below the initial yield strength. The fatigue limit for hard rocks when many compressive stresses are applied is about 60–75 per cent of initial compressive strength; for repeated tensional stresses the figure is about 50–60 per cent; and for alternating tensional and compressive stresses it is about 30 per cent.

Fatigue failure in hard rocks probably occurs by microcrack extension as strain energy is stored. The results of some experimental work are that the stresses associated with a large number of cycles of temperature change (20 000–25 000) may lead to microcrack development.

Plastic failure

This is a non-recoverable strain, characterised by slow deformation rather than catastrophic failure, and typical of less-brittle rocks. It is illustrated by the behaviour of certain clay minerals (see below) when they absorb water and expand. The internal changes that creep involves may alter a rock's resistance to weathering.

READINGS FOR FURTHER STUDY

In this chapter we have focused on the physical properties of rocks that encourage or resist the weathering processes. You will find much useful material in geological textbooks, but it needs to be hunted down. It is a good idea to look at the chapter headings and indexes of several books to find what you want.

The list below contains more specific readings.

Gerrard, A.J. 1988: *Rocks and landforms*. London: Unwin/Hyman. Considers rock properties from a geomorphological viewpoint.

Hochella, M.F. and White, A.F. (eds) 1990: *Mineral–water interface geochemistry. Reviews in Mineralogy*, **23**. An advanced book, so use with care. You will find material on the nature of mineral surfaces at the smallest scale.

McLean, A.C. and Gribble, C.D. 1985: *Geology for civil engineers*. London: Allen and Unwin. Will give you another perspective on many of the ideas about the physical properties of rocks.

Pettijohn, F.J. 1975: The texture of sediments. In *Sedimentary rocks*. New York: Harper and Row. Elaborates upon the ideas of packing, porosity and permeability.

West, G. 1991: *The field description of engineering soils and rocks*. Geological Society of London. Professional Handbook. Milton Keynes: Open University Press. Includes information on the description of soils and soft rocks, and on the measurement of rock and soil strength.

4

PHYSICAL PROPERTIES OF WATER

Because water is so widely distributed and because it has a variety of distinctive and important properties, it plays a dominating role in most weathering processes. It may function as the following.

- **A simple solvent** in which substances are dissolved; for example, Searle's Lake in California is alkaline due to the presence of dissolved sodium carbonate.
- **An active component** in a chemical reaction; for example, the hydrolysis of silicates such as occurs on the exposed outcrops of granite on Dartmoor. It has even been claimed that due to the prevailing winds causing preferential weathering and thus rounding of Dartmoor tors that these can be used as a compass!
- **A source of physical force** consequent upon freezing; for example, the extensive screes of upland Snowdonia in Wales produced by frost shattering at the end of the last Ice Age and possibly earlier.

As water contributes to many weathering processes, an understanding of its nature and properties is necessary in order that the processes themselves can be properly understood. In this chapter these ideas and interrelationships are developed by a detailed treatment of the physical properties of water (but note that surface tension and capillarity are discussed in Chapter 3).

Many of the properties of water are very familiar but others, e.g. dipole moment and dielectric constant, are rarely discussed in detail in text books on weathering, but nonetheless are crucial to an understanding of the behaviour of water, particularly in its ability to function as a solvent or as a medium in which chemical reactions of weathering may occur. Many of the properties can be explained from a knowledge of the structure of the water molecule, and therefore this will be treated first.

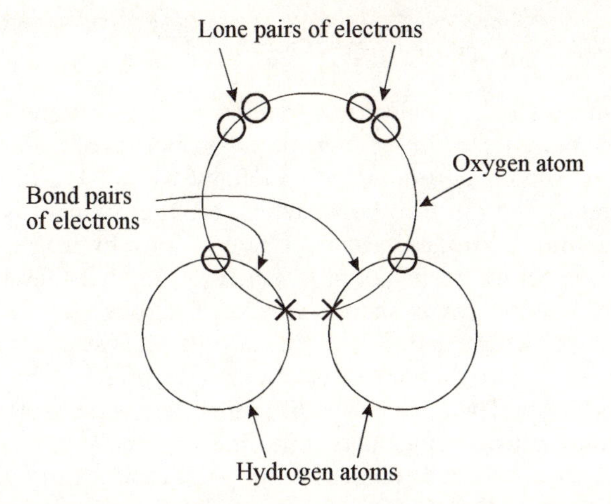

FIGURE 4.1 The electronic arrangement in the water molecule. × = electron originally of hydrogen; ○ = electron originally of oxygen.

4.1 STRUCTURE OF THE WATER MOLECULE

Water consists of a collection of water molecules, each of which has the formula H_2O and so is made up of two hydrogen atoms and one oxygen atom arranged in an angular fashion. This was discussed in Chapter 1 (*see* Fig. 1.11). The oxygen atom forms two covalent bonds, one to each hydrogen atom. Remember that covalent bonds arise from the sharing of electrons between atoms. In the case of the hydrogen atoms and oxygen atoms needed to form molecules of water each hydrogen atom has one electron and each oxygen atom has six outermost electrons. The sharing of electrons between the atoms is shown in Fig. 4.1.

It should be noted that there are two bond pairs of electrons, each of which consists of a shared pair of electrons with one provided by the hydrogen atom and one provided by the oxygen atom. This leaves four non-bonding electrons on the oxygen atom and these are arranged in pairs (lone pairs). A more accurate representation of the water molecule is shown in Fig. 4.2 in which the orientation of the pairs of electrons towards the corners of a regular tetrahedron is shown.

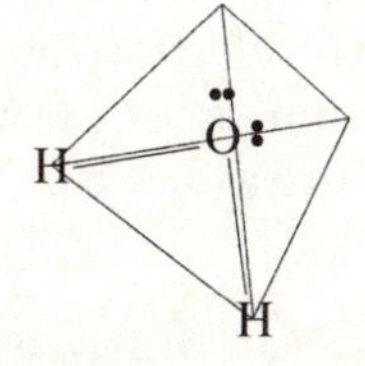

FIGURE 4.2 The relative orientation of pairs of electrons in the water molecule.

4.2 Dipole moment

The orientation of pairs of electrons in the water molecule will later be seen to be important in determining some of the properties of water, but if we are to understand the property of dipole moment we must also investigate the nature of the shared electron pairs, which are the covalent bonds between the oxygen and hydrogen atoms. Oxygen and hydrogen atoms have different values of electronegativity, which means that they have different powers of attraction for a shared pair of electrons (*see* Chapter 1). The electronegativity of oxygen, 3.5, is the second highest of all the elements. In contrast, hydrogen has an electronegativity of 2.1, which is near the middle of the scale. The fairly large difference between the electronegativity of oxygen and hydrogen means that in the water molecule, oxygen attracts the electrons of the shared covalent bond pairs quite strongly to itself.

This can be thought of in the following way: isolated atoms have zero charge, but as soon as they bond together, the atoms compete for the shared pairs of electrons in the bond. The atom which has the higher electronegativity will attract the bond pair more. This means that the shared pair is held closer overall to the oxygen atom than to the hydrogen atom. Thus the O–H bond in the water molecule is *polar*, which is shown as

$$\overset{\delta^-}{\mathrm{O}}-\overset{\delta^+}{\mathrm{H}}$$

where δ^- denotes an effective small negative charge on the oxygen atom and δ^+ denotes an effective small positive charge on the hydrogen atom.

The unequal sharing of electrons in the bonds and the angular structure of the water molecule has the important consequence of giving water a *dipole moment*, with a value of 1.85 debye (*see* Box 4.1). It should be noted that it is the combination of polar bonds with angular structure which gives

Box 4.1 Measurement of dipole moment

Dipole moments are measured in debye units (D) which are named after the Dutch chemist who developed ideas about dipole moment. An electric dipole results from equal but opposite charges next to each other. The modern definition relies on a charge of +1 coulomb (the unit of electric charge, C) separated from a charge of −1 coulomb by a distance of 1 metre, so it is the coulomb-metre. The original debye unit was based on the charge on an electron (negative) separated by 100 picometres ($1\,\mathrm{pm} = 10^{-12}\,\mathrm{m}$) from an equal and opposite (positive) charge. This gave a dipole moment of 4.8 D. The debye can be converted to SI units as follows:

$$1\,\mathrm{D} = 3.336 \times 10^{-30}\,\mathrm{C\,m}$$

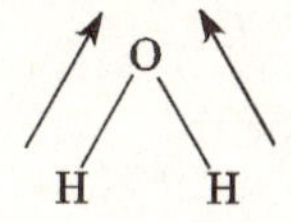

FIGURE 4.3 The angular structure of the water molecule and associated bond polarities.

water its dipole moment, but that there are other molecules which have polar bonds, but do not have a dipole moment. For example, you should now think about the geometric arrangement of the atoms of the carbon dioxide molecule, which has a linear structure:

$$O{=}C{=}O$$

The electrons in the C=O bonds are unequally shared and so the C=O bonds will be polar:

$$\overset{\delta^+}{C}{=}\overset{\delta^-}{O}$$

but overall the bond polarities cancel each other out. This can be seen when the bond polarities are represented by an arrow whose length indicates the magnitude of the polarity and whose direction shows where the more negative end occurs:

$$\underset{\longleftarrow\;\longrightarrow}{O{=}C{=}O}$$

This molecule therefore has polar bonds, but because of their relative orientation they cancel each other out, so that the molecule has a zero dipole moment.

In contrast, the polar bonds within the water molecule are arranged in an angular fashion and consequently do not cancel each other out, making the water molecule overall polar with a measurable dipole moment as shown in Fig. 4.3. So we can have

- **polar bonds** where there is unequal sharing of electrons in covalent bonds (note that non-polar covalent bonds occur in molecules such as H_2 and Cl_2 where each atom has an equal attraction for the electrons);
- **polar molecules** which have a dipole moment greater than zero (note that non-polar molecules, e.g. carbon dioxide, discussed above may contain polar bonds).

4.3 HYDRATION ENERGY

It has been shown that water is polar and therefore it might be expected to interact with any charged materials that are present. The importance of this type of interaction may be illustrated by the solution of ionic solids in water.

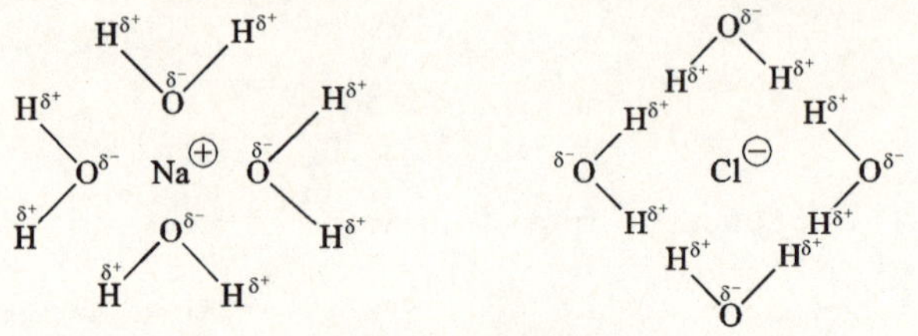

FIGURE 4.4 The interaction of water molecules with ions in solution.

You should remember that the dissolving material is called the *solute* and the liquid in which it dissolves is called the *solvent*.

Solution occurs in semi-arid regions, when occasional rain storms bring about the dissolution (dissolving) of available salts (ionic solids). Subsequent evaporation means that salts are then precipitated and are not removed from the area. In more humid regions the salts would be exported in solution.

When rock salt (NaCl) dissolves in water, both sodium ions, Na^+, and chloride ions, Cl^- are present in the aqueous solution. Taking each in turn, for Na^+ the polar water molecule will have its negative (oxygen) end attracted to the positively charged Na^+, leading to an arrangement in which six water molecules are quite strongly attracted to the Na^+. This is called *hydration of the ion* and is shown in Fig. 4.4. The chloride ion is also hydrated by water molecules, but in this case it is the positive hydrogen end of the water molecule which is attracted to the chloride ion (*see* Fig. 4.4). You can now understand how ions are carried away in solution from a weathering site.

Hydration energy, which is more correctly called the enthalpy of hydration (*see* Chapter 1), can be defined as the amount of energy liberated when 1 mole of gaseous ions is passed into water. This can be summarised for sodium ions as follows:

$$Na^+(g) \longrightarrow Na^+(aq)$$

and for chloride:

$$Cl^-(g) \longrightarrow Cl^-(aq)$$

The strength of interaction of ions and water molecules is an important factor in determining the ease with which ions can be extracted from a weathering substance by water molecules. Values of hydration enthalpy are given in Table 4.1. The value of hydration enthalpy is related to the relative size and charges of ions. The smaller the size of the ion, the more strongly water molecules are attracted to it, because the charge density at the surface of the ions is greater when they are smaller. Thus the smaller Na^+ ion has a hydration enthalpy some 84 kJ mol^{-1} greater than the larger K^+. Similarly, the doubly charged Ca^{2+} has a hydration enthalpy more than three times greater than that of the singly charged Na^+.

TABLE 4.1 Hydration enthalpy of some ions[a]

Ion	Hydration enthalpy ($kJ\,mol^{-1}$)[b]
Na^+	−406
K^+	−322
Ca^{2+}	−1577
Cl^-	−381
F^-	−515

[a] The process is gaseous ion → hydrated ion.
[b] A negative value indicates that energy is given out and the process is energetically favoured.

4.4 SOLUBILITY

The solubility of ionic solids in water is of great importance in weathering because of the widespread occurrence of water and its frequent contact with earth materials. Even the leaching of ions out of silicate minerals is an example of the phenomenon of solubility in action.

Solubility is a complex subject. We can begin by considering the energy factors which are important for solubility. As we saw in Chapter 1, the lattice enthalpy of an ionic solid is the energy liberated when an ionic solid is formed. If the solid is to be broken down so that its ions are separated, there must be some compensation for the loss of lattice enthalpy. This compensation comes from the energy released when the ions are hydrated, i.e. hydration enthalpy. We can express this as a simple equation:

$$\text{enthalpy of solution} = \text{hydration enthalpy} - \text{lattice enthalpy}$$

A negative value of enthalpy of solution would represent energy given out and thus be favourable to the process of solution. We can calculate the enthalpy of solution of NaCl using values of lattice enthalpy from Table 1.2 and hydration enthalpies from Table 4.1.

$$\begin{aligned}\text{enthalpy of solution of NaCl} &= (\text{hydration enthalpy of } Na^+ \\ &\quad + \text{hydration enthalpy of } Cl^-) \\ &\quad - \text{lattice enthalpy of NaCl} \\ &= -406 + (-381) - (-781) \\ &= -787 + 781 \\ &= -6\,kJ\,mol^{-1}\end{aligned}$$

We would predict that sodium chloride (NaCl) would be soluble in water since the enthalpy of solution is negative. A similar calculation for potassium chloride (KCl) leads to an enthalpy of solution of $+7\,kJ\,mol^{-1}$. The solubility of sodium chloride at 25°C is 36 g per 100 g of water and of potassium chloride is 35 g per 100 g of water. The solubilities are very similar, and this indicates that values of enthalpy of solution near to zero may correspond to a soluble ionic solid. The enthalpy of solution of calcium fluoride (CaF_2), which occurs in nature as the mineral fluorite, has an enthalpy of solution of $+27\,kJ\,mol^{-1}$, so we would expect a low solubility, which indeed it has being 0.0016 g per 100 g of water. In other words calcium fluoride is virtually insoluble. Notice that enthalpies of solution are obtained by taking the difference between two relatively large enthalpies (hydration and lattice) and that these enthalpies of solution are relatively small, so that small changes in hydration and lattice enthalpies can have a large effect on enthalpy of solution. The lattice and hydration effects tend to work in the same direction. A small, highly charged positive ion will tend to give a large lattice enthalpy because it attracts negative ions more strongly. Also, it will tend to attract the negative (oxygen) end of the water molecule and so give a greater hydration enthalpy.

In addition, the energetics of solution of an ionic solid are governed not just by the enthalpy change which occurs, but also by a factor called *entropy*. This is a measure of the randomness in a system, and is discussed further in Chapter 7, but here we can note that the greater the degree of randomness, the greater is the entropy. For example, a gas whose molecules move relatively freely has a higher entropy than a liquid. An increase in entropy favours a process, but for many processes where the enthalpy changes are of the order of 100 kJ (e.g. chemical reactions) it will be the enthalpy changes which are dominant. Because enthalpy changes associated with solubility are usually small, entropy changes may be important.

When a solid dissolves in water you might expect entropy to increase because the highly ordered ionic solid is broken down and the ions are spread through the liquid. There is, however, another effect, which is that the water molecules are no longer all freely moving as some are held by the ions. This means that entropy of solution may be positive (i.e. it decreases) or negative (i.e. it increases). Usually it is positive for ions with single charges, but with ions of higher charge it may be negative because the more highly charged ions have a greater influence on the water molecules.

To summarise:

- ionic solids will have appreciable solubility if the enthalpy of solution is negative or near to zero;
- ionic solids will be virtually insoluble if the enthalpy of solution is appreciably positive (e.g. $+27\,kJ\,mol^{-1}$ for CaF_2);
- entropy factors may be important and will cause an increase in solubility of ionic solids in water if the ions have single charges.

You can use these ideas to guide your thinking about weathering by the leaching of ions from, for example, silicates and aluminosilicates. Highly charged positive ions, e.g. Fe^{3+}, will be held more strongly in the solid silicate, but also will have a greater tendency to form a hydrated ion in solution, e.g. $Fe(H_2O)_6^{3+}$.

The dominant effect is normally that highly charged positive ions will tend to be retained in structures. Singly charged positive ions (e.g. Na^+)

are more easily removed in solution in water ('leached'). The tendency for an ion to be bound in to the solid is the dominant effect for highly charged ions.

4.5 DIELECTRIC CONSTANT

In the previous section we calculated enthalpies of solution for particular ionic solids. We know that many ionic solids are soluble in water. The inherent ability of water to be a good solvent is shown by its high value of dielectric constant which measures the ability to reduce forces between ions. This can be shown by

$$F = \frac{q^+q^-}{4\pi\varepsilon r^2}$$

where F is the force of attraction between ions
q^+ and q^- are the charges on the positive and negative ions respectively
ε is the dielectric constant (ε is the Greek letter epsilon)
r is the distance of separation of the ions.

For a vacuum the dielectric constant is called ε_0. For solvents the dielectric constant is usually expressed as $\varepsilon/\varepsilon_0$, i.e. it is the ability of the solvent to reduce forces between ions in the solvent as opposed to in a vacuum.

For pure water the dielectric constant ($=\varepsilon/\varepsilon_0$) is 82. The value of 82 for the dielectric constant means that the forces of attraction between ions in the water are reduced to about 1 per cent (strictly 1/82) compared to the forces between the same ions in a vacuum. This high figure explains why water is so good at overcoming the forces of attraction of ions in an ionic solid, resulting in the formation of aqueous solutions.

4.6 SOLUBILITY OF GASES IN WATER

As well as dissolving ionic solids, water can also dissolve gases, e.g. carbon dioxide (CO_2), oxygen (O_2), sulphur dioxide (SO_2) and ammonia (NH_3). This can have important weathering consequences; for example, dissolved carbon dioxide in water makes it acidic, and dissolved oxygen plays an important role in biological weathering.

The solubility of gases in water is described by Henry's law which relates the solubility of a gas in a liquid to the partial pressure of the gas above the liquid. The partial pressure of a gas comes from its proportion in a mixture of gases. Henry's law constant (K_H) is equal to the molar concentration at equilibrium of the gas in the liquid divided by the partial pressure of the gas in the gas phase at a particular temperature. For example, for oxygen (O_2) in water which is in contact with air:

$$K_H = \frac{[O_2(aq)]}{P_{O_2}}$$

Box 4.2 Calculation of the solubility of oxygen from air in contact with water

The Henry's law constant for oxygen at 20°C is

$$1.3 \times 10^{-3}\ \mathrm{mol\,L^{-1}\,atm^{-1}}$$

The partial pressure of oxygen in air is 0.21 atm:

$$K_H = \frac{[O_2(aq)]}{P_{O_2}}$$

$$1.3 \times 10^{-3} = \frac{[O_2(aq)]}{0.21}$$

$$\begin{aligned}[O_2(aq)] &= 1.3 \times 10^{-3} \times 0.21 \\ &= 2.73 \times 10^{-4}\ \mathrm{mol\,L^{-1}} \\ &= 2.73 \times 10^{-4} \times 32 \quad \text{(32 is the RMM of } O_2\text{)} \\ &= 0.0087\ \mathrm{g\,L^{-1}}\end{aligned}$$

where $[O_2(aq)]$ is the concentration in mol dm^{-3} of O_2 in water (the squared brackets indicate that this refers to concentration), and P_{O_2} is the partial pressure of the O_2 in air (=0.21 atmosphere).

We can rearrange this equation:

$$[O_2(aq)] = K_H P_{O_2}$$

Therefore

$$[O_2(aq)] \propto P_{O_2}$$

So the concentration of O_2 in water is proportional to the pressure of O_2 in contact with the water. It should be remembered that this is an idealised situation since often equilibrium will not be established and the concentration of the gas will be less than the equilibrium value. An example of a calculation of concentration of O_2 in water in contact with air is shown in Box 4.2.

Table 4.2 Solubility of pure gases in water[a] at various temperatures

Gas	Solubility (g kg^{-1} of water)			
	0°C	**20°C**	**25°C**	**30°C**
CO_2	3.35	1.69	1.45	1.26
O_2	0.069	0.043	0.039	0.036
SO_2	228	113	94	78
NH_3	897	529	480	410

[a] All figures are quoted for total pressure of gas and water of 1 atm.

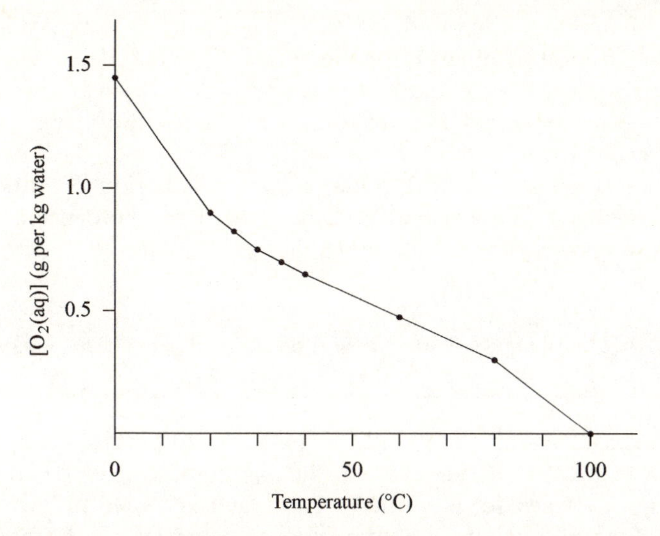

FIGURE 4.5 Variation of solubility of oxygen from air in contact with water.

The solubility of gases in liquids is quite dramatically affected by temperature (*see* Table 4.2). The solubility of oxygen in water in contact with air is plotted in Fig. 4.5. The solubility falls by almost 40 per cent between 0°C and 20°C. This variation in solubility with temperature may influence weathering involving oxygen since the concentration of dissolved gases in water in, for example, polar regions is potentially higher than that in tropical regions. The variation in solubility with temperature is generally true for all gases and is of particular significance for carbon dioxide. In cold environments the high solubility of this gas may offset the slower rates of reaction which occur with lower temperatures.

4.7 DIFFUSION OF OXYGEN IN SOILS

In biological weathering processes, e.g. in soil, oxygen is often important particularly to the function of decomposers. These are bacteria that break down organic matter in the soil.

As this oxygen is used up it needs to be replaced by oxygen from the atmosphere. The rate of diffusion is given by Fick's law:

$$Q = D\frac{\mathrm{d}c}{\mathrm{d}z}$$

where Q is the diffusion rate of the gas (kg m^{-2} s^{-1}), D is the *diffusion coefficient* in soil (m^2 s^{-1}), c is the concentration of gas in soil air (kg m^{-3}) and z is the depth in soil (m). $\mathrm{d}c/\mathrm{d}z$ is the expression for the change of concentration with depth. The use of 'd' is the standard mathematical way of showing a very small change and will be familiar to those of you who have studied calculus.

If soil is aerated, then soil air will usually have approximately the same concentration as air in the atmosphere. This is because D_{air} is large ($2 \times 10^{-5}\ m^2\ s^{-1}$). If the soil has a high water content, then anaerobic conditions may occur because D_{water}, the diffusion coefficient of oxygen in water, is much lower ($2 \times 10^{-9}\ m^2\ s^{-1}$).

You should appreciate that under actual weathering conditions, biological processes in the soil may considerably affect the concentrations of various gases present in the soil atmosphere.

4.8 Freezing and behaviour at temperatures near 0°C

When temperatures fall to around 0°C there are important changes in the behaviour of water. These have particular significance for mechanical weathering processes. Ice is formed when water is cooled to 0°C or below. This new phase involves a remarkable transformation. Each molecule becomes linked by hydrogen bonding to its four nearest neighbours to form a tetrahedral group (*see* Fig. 4.6a). These groups are in turn linked by hydrogen bonds to form a rigid hexagonal (i.e. six-sided) structure (*see* Fig. 4.6b). This organisation has two important consequences.

- As ice forms there is an increase in volume of about 9 per cent. This has great significance for the frost weathering process (*see* Chapter 3).
- If an outside force is applied to this ordered structure it will bring about collapse, i.e. change to water. This explains the observation that pressure may cause ice to melt at temperatures below freezing, e.g. at 1000 atm the

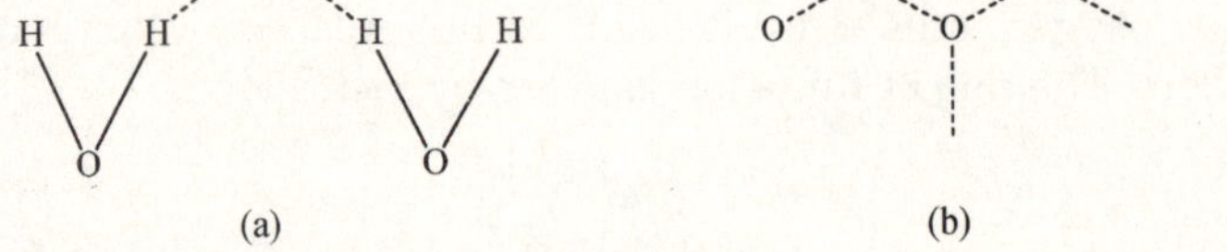

(a)

(b)

Figure 4.6 The structure of ice. (a) A two-dimensional molecular group, showing the four hydrogen bonds (– –). (b) A two-dimensional plan view of the open hexagonal arrangement of ice. Each O represents an oxygen which is attached to four hydrogen atoms as shown in (a). The structure is joined together by hydrogen bonds (– –). Note that the full structure in three dimensions involves a fourth bond to other hexagonal arrangements (not shown).

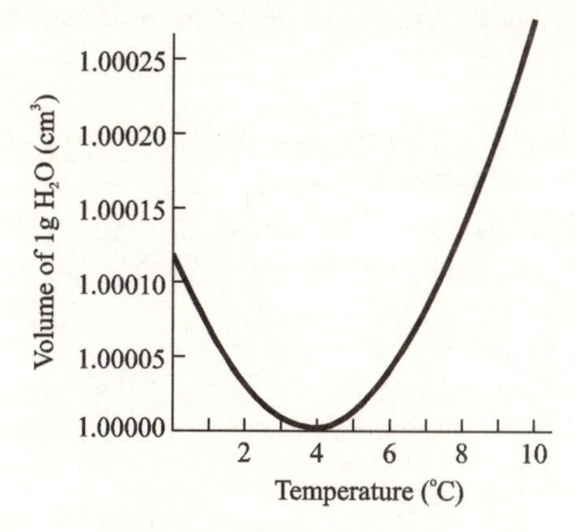

FIGURE 4.7 Variation in volume of water with temperature.

TABLE 4.3 Density of water and ice

Temp (°C)	Density ($g\,cm^{-3}$)
−10	0.9981 (liquid)
	0.9182 (solid)
0	0.9998 (liquid)
	0.9168 (solid)
5	1.0000
10	0.9997
15	0.9991
20	0.9982
25	0.9970
30	0.9956

melting point falls to −5°C. This pressure has implications for the growth of ice in rock cracks and pores (*see* Chapter 6).

Apart from pressure an increase in temperature will also cause ice to melt. The hydrogen bonds begin to break and the structure starts to collapse. The mobile water molecules are packed together more closely because of the loss of hydrogen bonds, volume is reduced and density is increased. Maximum density is reached at about 4°C (*see* Fig. 4.7). Variation in density changes with temperature between −10°C and 30°C is shown in Table 4.3. As temperature rises, the thermal motion of the molecules increases. This motion dominates over the diminished amount of hydrogen bonding and so volume increases.

READINGS FOR FURTHER STUDY

This topic is covered well in many general chemistry text books written for first year undergraduate courses. A text with very clear explanations is:

Atkins, P. and Jones, L. 1997: *Chemistry, molecules, matter and change,* 3rd edn. New York: Freeman.

The earlier texts involving authorship by P.W. Atkins on *General chemistry* also contain clear explanations:

Atkins, P.W. and Beran, J.A. 1990: *General chemistry,* 2nd edn. New York: Scientific American Books (W.H. Freeman).

5

CHEMICAL PROPERTIES OF WATER

In the previous chapter we saw that water has physical properties which are significant in weathering processes. In this chapter you will find that the chemical properties of water are also important in weathering. Water is at the heart of ideas about acidity and basicity, which are important features of chemical weathering. It is also the medium in which most reduction and oxidation weathering reactions occur.

You should note, however, that some reactions also occur out of solution, e.g. the oxidation of oxidisable solids by oxygen in the air. The general principles of reduction and oxidation are introduced in this chapter together with associated ideas of energy changes in these reactions. Finally, natural aqueous systems are discussed.

5.1 Acidity, basicity and pH

Water can undergo an important process called self-ionisation, in which one of the bonds of a water molecule breaks to give rise to a hydrogen ion (H^+) and a hydroxide ion (OH^-):

$$H_2O(l) \rightleftharpoons H^+(aq) + OH^-(aq)$$

where (l) = liquid and (aq) = aqueous (these describe the state of the species).

When the bond breaks in the water molecule, the pair of electrons in the bond remains with the oxygen of the OH^- group. The hydrogen therefore has no electrons and so it has a positive charge (H^+), while the hydroxide group has one extra electron because it retains both electrons of the bond (rather than the one which it originally contributed) and so has a negative charge.

The hydrogen ion (H^+) is not found separately in nature since it always combines with a water molecule to form the H_3O^+ ion. This may be called either the *hydronium ion* or the *oxonium ion*.

The self-ionisation of water only occurs to a very limited extent; for example, for pure water at 25°C (298 K) the concentration of H_3O^+ is only 10^{-7} mol dm^{-3}. The longer arrow going to water in the equation indicates that the equilibrium lies well to the left, so most of the water is non-ionised. You may remember that hydrogen has a relative atomic mass which is approximately 1, so 1 mole of H^+ will have a mass of 1 g. A concentration of 10^{-7} mol dm^{-3} means that 1 litre of water (1 dm^3) contains one ten-millionth of a gram of H^+!

Similarly, the concentration of OH^- is 10^{-7} mol dm^{-3} in pure water. So for pure water at 25°C,

$$[H_3O^+] = [OH^-] = 10^{-7}\ \text{mol dm}^{-3}$$

(remember that squared brackets indicate concentration of the species inside the brackets).

The very low concentration of hydrogen ions in water means that it is inconvenient to use these figures directly and so the pH scale is used (*see* Box 5.1).

Box 5.1 The pH scale and natural waters

pH is defined by:

$$\text{pH} = -\log_{10}[H_3O^+]$$

Thus for pure water at 298 K,

$$\begin{aligned}\text{pH} &= -\log_{10} 10^{-7} \\ &= -(-7) = 7\end{aligned}$$

By taking the log of the hydrogen ion concentration we produce a more easily managed number which is made positive by putting a minus sign in front of the logarithm. The pH scale usually runs from 0 to 14, but for most naturally occurring waters the values are fairly close to the middle of the scale.

- Oceanic waters have an average pH of 8.0–8.3, mainly due to dissolved calcium carbonate.
- Rain water, an important source of weathering solutions, may have a pH of 5.6 due to dissolved carbon dioxide.
- River and lake water varies from values above 7 down to values below 5, although the lower values of 5 or below have usually developed since the nineteenth century and arise from increased acidic air pollution and 'acid rain' from industrial development. This acid rain has created anthropogenic (human-generated) weathering effects, for example, on buildings constructed of limestone, since the limestone is basic and so reacts with the 'acid' rain to produce salts which may be washed away.
- The water in contact with peat, which is acidic, has a pH of 3–4.

The variation in concentrations of hydrogen ion ($[H^+]$) and thus pH can be understood by considering the self-ionisation of water mentioned earlier:

$$H_2O(l) \rightleftharpoons H^+(aq) + OH^-(aq)$$

Any substance which increases the concentration of H^+ will cause the pH to fall and is called an acid.

An acid is a proton (H^+) donor.

Similarly any substance which causes a decrease in the concentration of H^+ is called a base.

A base is a proton (H^+) acceptor.

The definitions arise from the Brønsted–Lowry theory developed in 1923. A more quantitative approach using the ideas of the law of mass action and equilibrium constant is described in Box 5.2.

The general principles of acids and bases can be shown by taking particular examples. If gaseous hydrogen chloride (HCl) is dissolved in water it is totally dissociated into H^+ and Cl^-:

$$HCl(g) + H_2O(l) \longrightarrow H_3O^+(aq) + Cl^-(aq)$$

Box 5.2 Quantitative basis of acidity and basicity – law of mass action and equilibrium constant

The *law of mass action* is based on the observation that for a chemical process there is an *equilibrium constant* (K_C) such that for a model reaction

$$A + B \rightleftharpoons C + D$$

at a particular temperature:

$$K_C = \frac{[C] \times [D]}{[A] \times [B]}$$

This means that when a reaction has come to equilibrium, there is a particular constant balance between the concentrations of the products and reagents. This is discussed further in Chapter 7. For the self-ionisation of water:

$$K_C = \frac{[H^+]\,[OH^-]}{[H_2O]}$$

Because the concentrations of H^+ and OH^- are very low we can write:

$$K_W = [H^+]\,[OH^-] = 10^{-14}$$

If we add a proton donor to water, the concentration of H^+ rises. In order to maintain the value of K_W constant, the equilibrium moves to the left so that the concentration of OH^- decreases. If we add a proton acceptor to water, e.g. a substance containing OH^-, then the concentration of OH^- increases, the equilibrium will move to the left and the concentration of H^+ decreases.

Box 5.3 The acidity of HCl

What is the pH of a solution containing 0.01 mol dm^{-3} of HCl?

HCl is totally dissociated, therefore the concentration of HCl is equal to the concentration of H^+:

$$[H^+] = 0.01$$
$$pH = -\log_{10}[0.01] = -\log_{10}[10^{-2}]$$
$$= 2$$

The HCl donates a proton to water, so it is an acid. You will find a quantitative example in Box 5.3.

Sodium hydroxide (NaOH) is a base and ionises when it is dissolved in water:

$$NaOH(s) \xrightarrow{\text{water}} Na^+(aq) + OH^-(aq)$$

The hydroxide ion is a proton acceptor as it can accept a hydrogen ion to form water (see quantitative example in Box 5.4).

For the reaction of typical acids and bases like HCl and NaOH we can write:

$$HCl(g) + H_2O(l) \longrightarrow H_3O^+(aq) + Cl^-(aq)$$
$$NaOH(s) \longrightarrow Na^+(aq) + OH^-(aq)$$
$$\underset{\text{from HCl}}{H_3O^+(aq)} + \underset{\text{from NaOH}}{OH^-(aq)} \longrightarrow H_2O(l)$$

Box 5.4 The basicity of NaOH

What is the pH of a solution containing 0.01 mol dm^{-3} of NaOH? Since

$$K_W = 10^{-14} = [H_3O^+][OH^-] \qquad (5.1)$$

and $[OH^-] = 0.01$ (i.e. 10^{-2}) mol dm^{-3}, in order to keep K_W constant we substitute the value of $[OH^-]$ in equation (5.1). So

$$10^{-14} = [H_3O^+] \times 10^{-2}$$

i.e.

$$[H_3O^+] = \frac{10^{-14}}{10^{-2}} = 10^{-12}$$

$$pH = -\log_{10}[H_3O^+] = -\log_{10} 10^{-12} = 12$$

You should notice that the presence of OH^- forces the self-ionisation of water to the left-hand side, thus lowering the concentration of the hydrogen ion.

Overall:

$$HCl + NaOH \longrightarrow NaCl + H_2O$$
$$\text{acid} + \text{base} \longrightarrow \text{salt} + \text{water}$$
$$(\text{salt} = \text{ionic solid})$$

We can summarise the use of the pH scale to define acidity and basicity:

$pH < 7$	Acidic
$pH = 7$	Neutral
$pH > 7$	Basic

This topic is discussed further in Chapter 7.

5.2 Redox reactions in water

Reduction and oxidation are concerned with the transfer of electrons (the symbol 'e' is often used for an electron) between different chemical species. It is helpful to remember 'OIL RIG':

*O*xidation *i*s *l*oss of electrons ('OIL')
*R*eduction *i*s *g*ain of electrons ('RIG')

As examples, consider the oxidation of Fe^{2+} to Fe^{3+}:

$$Fe^{2+} \longrightarrow Fe^{3+} + e$$

and the reduction of Fe^{3+} to Fe^{2+}:

$$Fe^{3+} + e \longrightarrow Fe^{2+}$$

These are of particular interest since both may occur during weathering. The oxidation of Fe^{2+} to Fe^{3+} occurs when Fe^{2+} is transported down through a soil profile to a region where alkaline oxidising conditions exist. Groundwater is often reducing, so any Fe^{3+} may be reduced and dissolved as Fe^{2+}, e.g. by reduction, with hydrogen sulphide produced by decomposition of organic material:

$$8Fe^{3+} + H_2S + 4H_2O \longrightarrow 8Fe^{2+} + SO_4^{2-} + 10H^+$$

If this groundwater emerges as a spring, the dissolving of oxygen from the air will cause the Fe^{2+} to be oxidised and produce the characteristic brown-red colour of ferric oxide (iron(III) oxide, Fe_2O_3).

5.2.1 Balancing equations

The term *redox reaction* means that one species in a chemical reaction is reduced ('red') while another is oxidised ('ox'). A simple example is the reaction between magnesium and oxygen:

$$Mg(s) + \tfrac{1}{2}O_2(g) \longrightarrow MgO(s)$$

In this reaction magnesium is oxidised:

$$Mg \longrightarrow Mg^{2+} + 2e$$

and oxygen is reduced:

$$\tfrac{1}{2}O_2 + 2e \longrightarrow O^{2-}$$

Notice that the magnesium has transferred two of its electrons to an oxygen atom to form O^{2-}. This allows us to *balance* the equation since one magnesium atom can only supply its two electrons to one oxygen atom – not two or three or more. We can illustrate this by looking at the reaction of sodium with oxygen. Sodium forms Na^+:

$$Na \longrightarrow Na^+ + e \qquad (5.2)$$

Oxygen forms O^{2-}:

$$\tfrac{1}{2}O_2 + 2e \longrightarrow O^{2-} \qquad (5.3)$$

Notice that sodium can only provide one electron but oxygen needs two, so two sodium atoms are needed to meet the requirements of the oxygen atom. We therefore multiply equation (5.2) by two:

$$2Na \longrightarrow 2Na^+ + 2e \qquad (5.4)$$

We can now add equations (5.3) and (5.4):

$$2Na + \tfrac{1}{2}O_2 + \cancel{2e} \longrightarrow 2Na^+ + O^{2-} + \cancel{2e}$$

Two sodium atoms have transferred two electrons to one oxygen to form $2Na^+$ and O^{2-}, which is Na_2O. The two electrons appear on either side of the equation and are cancelled out to give the overall balanced equation.

Another example is the reaction of sodium and chlorine:

$$Na \longrightarrow Na^+ + e \qquad (5.5)$$

$$\tfrac{1}{2}Cl_2 + e \longrightarrow Cl^- \qquad (5.6)$$

Here one chlorine ($\tfrac{1}{2}Cl_2$) requires one electron and a sodium can lose one electron, so we can add equations (5.5) and (5.6):

$$Na + \tfrac{1}{2}Cl_2 \longrightarrow Na^+ + Cl^- \quad (NaCl)$$

In this equation the sodium is oxidised and so the chlorine is called an *oxidising agent*. At the same time the chlorine is reduced and so the sodium is called a *reducing agent*.

5.2.2 Redox and the periodic table

You may have noticed in the examples above that the species which are oxidised are metallic (Na and Mg), while the species which are reduced are non-metallic (O_2 and Cl_2). Metallic elements, particularly those to the left-hand side of the periodic table in groups I and II as well as the heavier elements of groups III (group 13) and IV (group 14), usually tend to lose electrons to form positive ions (*cations*), e.g.

Na^+, K^+ (group I)
$Mg^{2+}, Ca^{2+}, Sr^{2+}, Ba^{2+}$ (group II)
Al^{3+} (group III)
Pb^{2+}, Pb^{4+} (group IV)

The non-metallic elements to the right-hand side of the periodic table, especially group VII (group 17; the halogens) and group VI (group 16), tend to gain electrons to form negative ions (*anions*), e.g.

F^-, Cl^-, Br^-, I^- (group VII)
O^{2-}, S^{2-} (group VI)

This just leaves the elements at the centre of the periodic table. These are the transition elements (they have properties intermediate between those of the left-hand side - reactive metals - and right-hand side - non-metals). They are all metals and tend to form cations, but they exhibit variable oxidation numbers (*see* Chapter 1). For example, we saw earlier that iron can form Fe^{2+} (ferrous iron or iron(II)) and Fe^{3+} (ferric iron or iron(III)). A very wide range of oxidation numbers may occur; for example, chromium can form Cr^{3+} (chromium(III)) and Cr^{6+} (chromium(VI) - in the dichromate ion, $(Cr_2O_7)^{2-}$, or the chromate ion, $(CrO_4)^{2-}$). Very high oxidation numbers such as 6 will not normally occur naturally, as an element with such an oxidation number will act as a strong oxidising agent and thus achieve a lower oxidation number.

5.2.3 Electromotive force

The strong tendency of chromium in a high oxidation number to form a lower oxidation number leads us to develop a scale of the relative ability to undergo reduction or oxidation. If we add some shiny white zinc metal to a blue solution of copper sulphate, we observe the precipitation of brownish red copper metal. This is a good visual illustration of the stronger tendency of zinc than of copper to form ions:

$$Zn + Cu^{2+} \longrightarrow Zn^{2+} + Cu$$

The competitive processes are:

$$Zn \longrightarrow Zn^{2+} + 2e$$

$$Cu \longrightarrow Cu^{2+} + 2e$$

The fact that Zn reduces Cu^{2+} to Cu while it is itself oxidised to Zn^{2+} shows that zinc is a stronger reducing agent than copper. Gold is a very weak reducing agent although it is usually regarded as an oxidising agent. This means while gold may be found in its oxidised form, e.g. gold sulphide,

Box 5.5 The definition of reduction potential

The term reduction potential is most commonly used, but different text books use different terms such as electrode potential, oxidation potential, redox potential. We have adopted the most widely used convention which is that the half reaction is written with the *reduced species on the right*, e.g.

$$Zn^{2+} + 2e \longrightarrow Zn^0$$

The electrons therefore appear on the left-hand side of the half-equation.

it is more commonly found as the pure metal, e.g. in the form of nuggets. A series can be constructed with the reducing agents listed from strongest to weakest, e.g.:

potassium (K) > calcium (Ca) > zinc (Zn) > copper (Cu) > gold (Au)

The relative tendency of electrons to be transferred to or from chemical species is measured as an electromotive force expressed as a reduction potential (*see* Box 5.5).

The standard reduction potential (E^0) for the hydrogen electrode is set at 0.00 V and other standard reduction potentials are quoted relative to the hydrogen value. The measurement of reduction potentials and the meaning of the word 'standard' are discussed in Box 5.6.

The values of reduction potentials may be negative or positive. *A negative value* indicates that the element has a stronger tendency to form ions than does hydrogen. It is easily oxidised and so is a reducing agent. For example,

Box 5.6 Measurement of standard reduction potentials

Measurements are made by dipping an electrode made of the metal under consideration in a solution of its own ions with a standard concentration of 1 mol dm^{-3} at a temperature of 25°C.

This is connected to a standard hydrogen electrode in which a platinum electrode dips into an acid solution with a concentration of H_3O^+ of 1 mol dm^{-3} in equilibrium with hydrogen (H_2) at a pressure of 1 atmosphere. The potential difference (in volts) is measured using a voltmeter. The value measured is called the standard reduction potential, E^0. It is 'standard' because of the defined conditions, i.e. concentrations of 1 mol dm^{-3}, pressure of gases at 1 atmosphere and temperature of 25°C. The value for the hydrogen electrode is set at 0.00 V.

consider

$$Zn^{2+} + 2e \longrightarrow Zn \qquad E^0 = -0.76\,V$$

Zinc is a stronger reducing agent than hydrogen:

$$H^+ + e \longrightarrow \tfrac{1}{2}H_2 \qquad E^0 = 0.00\,V$$

So zinc reduces H_3O^+ to H_2:

$$Zn + 2H^+ \longrightarrow Zn^{2+} + H_2$$

A positive value indicates that the element has a lesser tendency to form ions than does hydrogen. It is easily reduced and is an oxidising agent, e.g.

$$Cu^{2+} + 2e \longrightarrow Cu \qquad E^0 = 0.16\,V$$

In this case copper ions (Cu^{2+}) can oxidise hydrogen (H_2) to H^+ (or hydrogen can reduce copper ions to copper):

$$Cu^{2+} + H_2 \longrightarrow Cu + 2H^+$$

A full set of reduction potentials is given in Table 5.1.

TABLE 5.1 Some reduction potentials

Name	Reaction	Standard reduction potential E^0 (V)
Potassium	$K^+ + e \longrightarrow K$	−2.94
Barium	$Ba^{2+} + 2e \longrightarrow Ba$	−2.91
Strontium	$Sr^{2+} + 2e \longrightarrow Sr$	−2.90
Calcium	$Ca^{2+} + 2e \longrightarrow Ca$	−2.87
Sodium	$Na^+ + e \longrightarrow Na$	−2.71
Magnesium	$Mg^{2+} + 2e \longrightarrow Mg$	−2.36
Aluminium	$Al^{3+} + 3e \longrightarrow Al$	−1.68
Manganese	$Mn^{2+} + 2e \longrightarrow Mn$	−1.18
Zinc	$Zn^{2+} + 2e \longrightarrow Zn$	−0.76
Chromium	$Cr^{3+} + 3e \longrightarrow Cr$	−0.74
Sulphide	$S + 2e \longrightarrow S^{2-}$	−0.57
Iron	$Fe^{2+} + 2e \longrightarrow Fe$	−0.44
Nickel	$Ni^{2+} + 2e \longrightarrow Ni$	−0.24
Lead	$Pb^{2+} + 2e \longrightarrow Pb$	−0.13
Hydrogen	$2H^+ + 2e \longrightarrow H_2$	0.00
Copper	$Cu^{2+} + 2e \longrightarrow Cu$	0.16
Iodide	$\frac{1}{2}I_2 + e \longrightarrow I^-$	0.54
Mercury	$Hg^{2+} + 2e \longrightarrow Hg$	0.85
Chloride	$\frac{1}{2}Cl_2 + e \longrightarrow Cl^-$	1.36
Gold	$Au^+ + e \longrightarrow Au$	1.69
Fluoride	$\frac{1}{2}F_2 + e \longrightarrow F^-$	2.89

Source: Aylward, G. and Findlay, T. 1994: *SI Chemical data*, 3rd edn. John Wiley.

Spontaneous reaction

The direction of spontaneous reaction between two half-equations can be found by use of the reduction potentials (e.g. from Table 5.1):

$$Na^+ + e \longrightarrow Na \qquad E^0 = 2.71\,V \tag{5.7}$$

$$\tfrac{1}{2}Cl_2 + e \longrightarrow Cl^- \qquad E^0 = 1.36\,V \tag{5.8}$$

What is the spontaneous reaction?

First, it is known that:

$$\Delta G^0_{reaction} = -nE^0_{reaction}F$$

where $\Delta G^0_{reaction}$ is the Gibbs free energy change for the reaction (*see* section 7.3), n is the number of electrons involved and F is the Faraday constant (−96 500 coulombs). For a spontaneous reaction ΔG^0 must be negative (*see* section 7.3). This means that $E^0_{reaction}$ must be positive.

We can combine equations (5.7) and (5.8) in two possible ways. First, reverse equation (5.7) and add equation (5.8) to it:

$$Na + Cl_2 \longrightarrow Na^+ + Cl^-$$

$$E^0_{reaction} = -\,(-2.71) + 1.36$$

$$= 4.07\,V$$

Second, reverse equation (5.8) and add equation (5.7) to it:

$$Na^+ + Cl \longrightarrow Na + \tfrac{1}{2}Cl_2$$

$$E^0_{reaction} = -2.71\,V - (+1.36)$$

$$= -4.07\,V$$

For the first alternative:

$$E^0_{reaction} = 4.07\,V$$

$$\Delta G^0_{reaction} = -\,1 \times 4.07 \times 96\,500$$

$$= -\,392\,755\ \text{joules}$$

The negative value means that this is the spontaneous reaction.

5.2.4 The Nernst equation, Eh and pH

The calculation of spontaneous reaction just carried out was for standard conditions, i.e. concentrations of $1\,mol\,dm^{-3}$, pressures of 1 atmosphere and a temperature of 25°C (298 K). The conditions under which weathering occurs will normally not be the standard ones so an expression which allows for this is required.

For a model reaction:

$$A + B \rightleftharpoons C + D$$

the Nernst equation is:

$$E = E^0 + \frac{2.303RT}{nF}\log\frac{[C][D]}{[A][B]}$$

at 25°C (298 K) we can simplify this by substituting

$$T = 298\,K$$

$$R = 8.314\,J\,K^{-1}\,mol^{-1}$$

$$F = 96\,485\,C\,mol^{-1}$$

so that

$$E = E^0 + \frac{0.059}{n}\log\frac{[C][D]}{[A][B]}$$

Notice that the Nernst equation contains a dependence on temperature and on the concentrations of reagents as well as products, so that we can calculate an electromotive force (E) for a reaction under non-standard conditions.

Under natural conditions there will usually be a mixture of species present and so *E from the above equation is replaced by Eh* in which the electromotive force of the natural water is measured against the standard hydrogen electrode. The same principles apply as earlier, so if Eh is negative then reducing agents are present, and if Eh is positive then oxidising agents are present.

Possible redox reactions in water are limited at one extreme by the possibility of water being oxidised to oxygen:

$$2H_2O \rightleftharpoons O_2 + 4H^+ + 4e \qquad E^0 = 1.23\,V$$

(i.e. O^{-II} is oxidised to O^0), and at the other extreme by the possibility of H^+ from H_2O being reduced:

$$H_2 \rightleftharpoons 2H^+ + 2e \qquad E^0 = 0.00\,V$$

Both of these extremes show a pH dependence which can be calculated (*see* Box 5.7 for the upper value and Box 5.8 for the lower value).

We can use the general expressions derived in Boxes 5.7 and 5.8 to calculate the upper and lower limits of redox stability in water for different values of pH. Some representative values are given in Table 5.2.

The range of Eh and pH values for some natural systems are shown in Fig. 5.1. At the extremes it can be seen that where water is in contact with the atmosphere, e.g. rain water, rivers, ocean surface water and mine waters, it is oxidising with positive Eh values. Waterlogged soils tend to be reducing with negative Eh values.

Three factors are important in determining Eh and pH values in natural environments:

- organic reactions, e.g. respiration (which gives CO_2), photosynthesis (gives O_2) and decay;

Box 5.7 The upper value of Eh – dependence on pH

The upper value of Eh depends on

$$2H_2O \rightleftharpoons O_2 + 4H^+ + 4e \qquad E^0 = 1.23\,V$$

$$Eh = 1.23 + \frac{0.059}{4}\log\frac{[H^+]^4 p_{O_2}}{[H_2O]^2}$$

We can ignore $[H_2O]^2$ since by convention this is set at 1.00. The partial pressure of oxygen (p_{O_2}) in the atmosphere is 0.21 atm.

Therefore

$$Eh = 1.23 + \frac{0.059}{4}\log[H^+]^4 \times 0.21$$

(Remember that $\log X^a \cdot Y^b$ can be rewritten as $a\log X + b\log Y$.)

Therefore

$$Eh = 1.23 + \frac{0.059}{4} \times 4\log[H^+] + \frac{0.059}{4}\log(0.21)$$

Now $pH = -\log[H^+]$. Therefore

$$\mathbf{Eh = 1.22 - 0.059\,pH}$$

Box 5.8 The lower value of Eh – dependence on pH

The lower value of Eh depends on

$$H_2 \rightleftharpoons 2H^+ + 2e \qquad E^0 = 0.00\,V$$

$$Eh = 0 + \frac{0.059}{2}\log\frac{[H^+]^2}{[H_2]}$$

(Remember $\log X^a/Y^b$ can be rewritten as $a\log X - b\log Y$.)

Therefore

$$Eh = \frac{0.059}{2} \times 2\log[H^+] - \frac{0.059}{2}\log p_{H_2}$$

$$= -0.059 \times pH - \frac{0.059}{2} \times \log p_{H_2}$$

The lowest value would occur when the partial pressure of hydrogen was 1 atmosphere and since $\log 1 = 0$ the equation is

$$\mathbf{Eh = -0.059\,pH}$$

TABLE 5.2 Eh and pH: redox stability limits for water

pH	Lower Eh (=−0.059 pH) (V)	Upper Eh (=1.22 − 0.059 pH) (V)
0	0	1.22
4	−0.236	0.984
8	−0.472	0.748
12	−0.708	0.512
14	−0.826	0.394

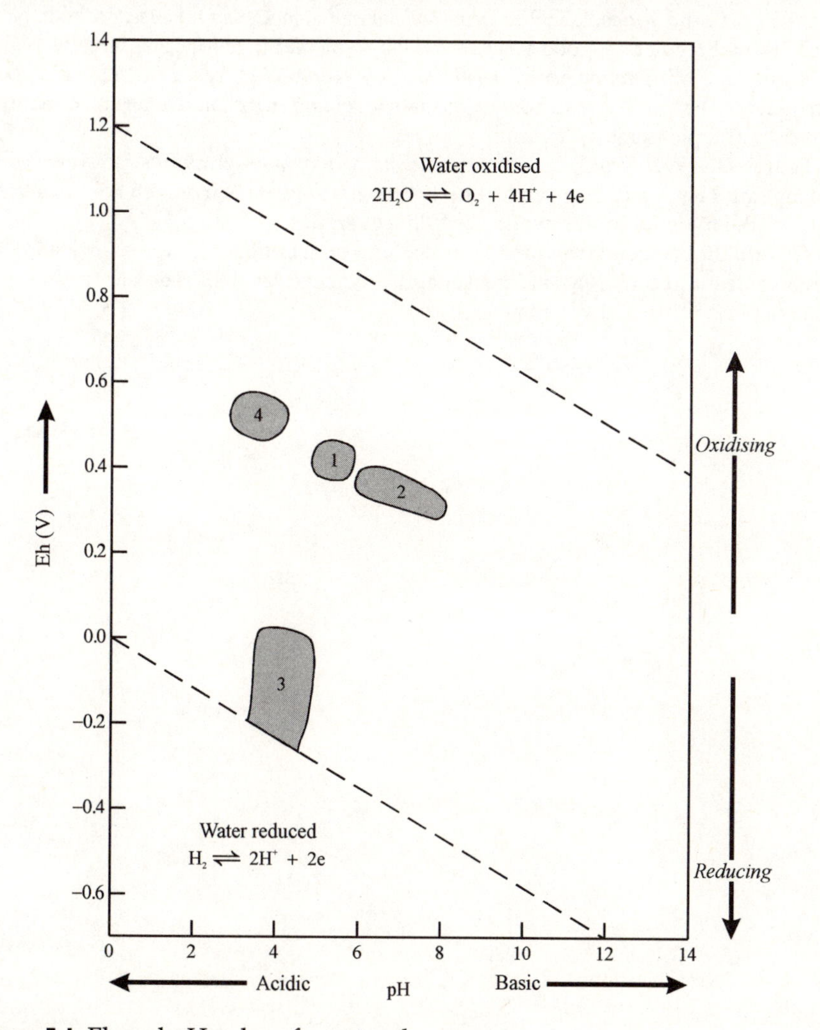

FIGURE 5.1 Eh and pH values for natural systems.
---- upper and lower limits of stability of water. 1. rain; 2. rivers; 3. surface ocean water; 4. waterlogged soils.

- redox reactions of common elements, e.g. iron, manganese and sulphur;
- carbonation (*see* section 7.11).

READINGS FOR FURTHER STUDY

The early material is quite straightforward and is covered in first-year general chemistry books such as those by Atkins. The later parts of the chapter are discussed, particularly in geochemical and environmental chemistry text books.

Atkins, P. and Jones, L. 1997: *Chemistry, molecules, matter and change*, 3rd edn. New York: W.H. Freeman. A good basic text book - see readings for Chapter 4.

Baird, C. 1995: *Environmental chemistry*. New York: W.H. Freeman. This is a good introductory text at first-year undergraduate level and has clear discussion of natural waters and topics such as acid rain.

Faure, G. 1991: *Principles and applications of inorganic geochemistry*. New York: Macmillan. This is an advanced text up to postgraduate level in which the theoretical basis of the material in this chapter is fully covered.

O'Neill, P. 1993: *Environmental chemistry*, 2nd edn. London: Chapman & Hall. This contains useful discussions, at a reasonably advanced level, of redox chemistry and other aspects of natural waters.

6

MECHANICAL WEATHERING PROCESSES

Mechanical (or physical) weathering processes involve the disintegration of rock with no chemical alteration. Their relationship with the chemical weathering processes is, however, close in that chemical alteration may reduce the strength of a rock to a level at which the stresses of mechanical weathering are sufficient to cause breakdown. It should be realised that there may be problems in identifying the exact contribution and order of each type of process.

In this chapter we will study six mechanical weathering processes:

- **low-temperature, water-based weathering**, which is made up of four distinctive activities: *freeze–thaw weathering, hydration shattering, ice crystal growth,* and *hydraulic pressure;*
- **salt weathering;**
- **wetting and drying;**
- **insolation weathering;**
- **pressure release;**
- **stress corrosion cracking.**

6.1 Low-temperature, water-based weathering

6.1.1 Introduction

The behaviour of water at temperatures around the freezing point gives rise to an effective process of rock breakdown. Investigators have used three main procedures in their study of low-temperature weathering, particularly freeze–thaw weathering. You will find it helpful to study these approaches before we look at the mechanisms themselves.

Laboratory studies

A widely used method has been to investigate freeze–thaw weathering by focusing on experiments that were supposed to be simulations of natural conditions of temperature and humidity. At the simplest level the approach involved placing rock samples in water, causing the temperature to fluctuate through 0°C, and then observing the results. You should note that such a procedure has several weaknesses.

- The moisture conditions during an experiment often differ significantly from those in nature. In particular, experimental conditions may exaggerate the amount of water occurring naturally.
- The rock samples used may not be representative. They are typically small (cubes of sides greater than 10 cm are rarely used) and so cannot be used for experiments simulating, e.g., large-scale joint-controlled failure.
- Little information emerges about the mechanics of failure. Other experiments have provided information about the behaviour of water at subzero temperatures and the likely pressures generated by freezing. Such information can be used to develop and test theoretical models of frost weathering.

Useful information about mechanisms has been provided by experiments on the reaction of building materials, such as concrete and natural stone, to low-temperature conditions.

Field studies

Many of the earlier field studies were based on an acceptance of the simple idea that the expansion by freezing of existing water caused rock destruction. Investigators collected information about the number and temperature range of freeze–thaw cycles, and the freezing rate, at and near rock surfaces. Later workers developed the approach by measuring subsurface temperatures through the use of devices such as thermistor sensors. These involve an electrical semiconductor in which resistance to current falls as temperature rises. Modern studies have focused on rock properties and on moisture availability at particular sites. The combination of more realistic field investigation and laboratory experiment has encouraged the development of more convincing theory.

Use of theoretical models

You should appreciate that theoretical models are needed in order to guide laboratory investigations and field studies, particularly by providing hypotheses for testing. Early theories of freeze–thaw weathering have not provided a complete explanation, and the emphasis is now on models that involve mathematical techniques and physical principles. An example is the model outlined in section 6.1.2 below, which combines mathematical manipulation, the principles of fracture mechanics, and laboratory evidence of water migration in freezing rock, to clarify the process of frost weathering.

In spite of advances, the processes of low-temperature, water-based weathering are still not completely understood. It is therefore not surprising that several theories are current, although you should not expect that only one theory should apply in all circumstances. We outline the main theories below, and look at the extent to which they are supported by theoretical, experimental and field observational work.

6.1.2 Freeze–thaw weathering

This is the process of rock disintegration that takes place when water freezes and so expands within rock. The process has also been called *frost weathering, frost shattering, frost wedging, frost (induced) cracking, frost bursting* and *gelifraction* ('breaking by freezing'). There is an important distinction between the freezing of in-place water and the freezing associated with water movement, and we will describe these two types separately.

Freeze–thaw weathering by the freezing of in-place water

Theory

This is the earliest, simplest and most natural idea. When water is frozen a 9 per cent volumetric expansion takes place. This is because the water molecules take up an open hexagonal arrangement which occupies more space than their relatively closely packed distribution in the liquid phase (*see* Chapter 4). The expansion generates pressure against the containing walls in a closed system. Under optimum conditions, at −22°C, a theoretical pressure of 207 MPa is developed. This pressure is much greater than rock tensile strength, which is rarely more than 10 MPa. In practice, however, this pressure would not be reached for several reasons.

- With increasing pressure the freezing point of water falls (*see* Chapter 4). Specifically, for each 1 MPa increase in pressure, the freezing point falls by 0.074°C, i.e. the 'pressure melting constant' for ice is $0.074°C\,MPa^{-1}$.
- The presence of salts reduces the freezing point.
- Air in rocks would be compressed as ice formed, thus absorbing some of the pressure.
- The presence of water suggests a point of entry and therefore that the system is open. Pressure would be released in the direction of water entry.
- The temperature of −22°C would rarely occur, other than in high-altitude and high-latitude environments.

However, high pressure associated with freezing could be maintained under certain conditions.

- Rapid freezing of water-filled cracks from the top downwards would allow ice to act as a seal. A freezing rate of more than $0.1°C\,min^{-1}$ may be necessary for sealing to occur. A consequence would be that pressures greater than 19.6 MPa could be generated at the bottom of a crack 1 mm wide and 100 mm deep in a saturated non-porous rock such as granite.

- The very rapid conversion of supercooled water (at about –5 to –6°C) to ice would prevent loss of pressure in the direction of water entry. The consequent breakdown of rock has been called 'frost bursting'.

Even if the theoretical pressure is not reached, you may think it is reasonable that the actual stresses are quite sufficient to shatter rock.

Contribution of laboratory and field evidence

The focus of early experimental work to test the theory was on the environmental conditions that might bring about rock breakdown. A typical experimental design was for a rock sample to be placed in water so that it was half-submerged. It was then subjected to a series of temperature cycles supposed to simulate those of major natural environments. Two commonly used cycles were the 'Icelandic', involving temperature variation between +8 and –8°C over about 24 h, and the 'Siberian', when the range was between +15 and –30°C over about 72 h. Investigators could not agree over which type of cycle was most effective for freeze–thaw weathering, but a minimum range seemed necessary.

Recent laboratory studies have focused on the 'threshold freezing intensity', i.e. the upper limit of rock temperature for frost shattering to occur. Investigators have suggested that this limit is about –3 to –5°C.

Temperatures recorded at field sites did not, however, seem to match those that the experimenters believed were necessary for rock disintegration. For example, the range of temperatures measured in bergschrunds in Norway and Switzerland in the summer months was small, varying from about +1.5 to –1.7°C. (A bergschrund is a crevasse (tensional crack) at the head or margin of a glacier. Bedrock may be exposed at its base which may be 45 m deep.) In the winter it may vary between –2 and –4.5°C, but little liquid water is then available. Similarly, measurements of temperatures on the rockwall above the headwall crevasse of a glacier in the Canadian Rockies showed a summer freeze–thaw environment but with a small range. At one site the mean daily maximum temperature was 2.5°C and the minimum –1.4°C.

In addition, the suggested freezing rate of 0.1°C min^{-1} has not been generally found. For example, a study of a bergschrund at the Jungfraujoch, Switzerland, showed a rate of temperature change in spring between 0.2 and 0.02°C h^{-1}, and an investigation of freezing rates during the summer on a snow patch in the Canadian Rockies yielded values of about 0.02°C min^{-1}.

The 'volumetric' hypothesis is simple and sounds convincing. It was assumed to be valid in many of the early field and laboratory investigations. You should, however, note that it is unlikely to be the main mechanism of freeze–thaw weathering for two reasons:

- the implied rapid rates and high amplitudes of temperature change have not been generally confirmed in the field;

- recent theory and experiment suggest that water is attracted to a freezing front rather than expelled from it, which would be a prediction from the volumetric hypothesis.

Freeze–thaw weathering by water migration and ice growth

Theory

This model of rock breakdown involves the movement of water to a freezing zone rather than its in-place conversion to ice. The model is therefore dynamic. It combines the various factors that modify the behaviour of freezing water, including:

- **rock properties** such as elastic moduli, tensional strength, grain size and shape, and crack size;
- **environmental conditions** such as temperature and its gradient, and water availability and behaviour.

A good starting point for studying this model is the observation that the two separate phases of ice and water can exist together in a rock at temperatures at and below 0°C (*see* Fig. 6.1). You should note that the concept of Gibbs free energy (*see* section 7.3) requires that two phases can only be stable together when their free energies are equal. If they are unequal, the phase of lower free energy (e.g. ice) will grow at the expense of the phase of higher free energy (e.g. water). As the evidence of Fig. 6.1 suggests, this

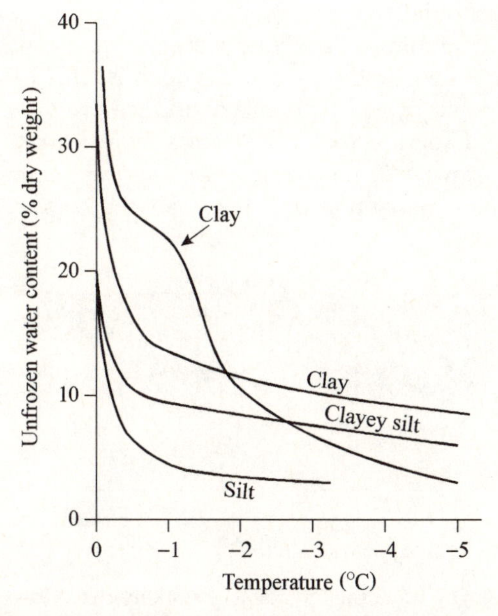

FIGURE 6.1 The liquid water content of two clays, a clayey silt and a silt, at various temperatures. (*Source*: Williams, P.J. 1982: *The surface of the Earth: an introduction to geotechnical science*. London: Longman Group, p. 96.)

may not happen, so one or more factors must be reducing the free energy of the water to about that of ice, and so effectively lowering the freezing point. In this way ice and water can coexist. These factors include the following.

- **The presence of salts**, which lowers the free energy and depresses the freezing point. The amount of depression is affected by the nature of the salt and its concentration, but is not very great, being about 1°C.
- **Adsorption effects**. The thin layer of polar water molecules adsorbed to a mineral surface is a zone of reduced free energy, i.e. lowered freezing point.
- **Capillary effects**. The high 'suction' generated in water at the top of a capillary column is associated with a lowered free energy and thus the freezing point.

You should now see why water and ice can exist together, but it is now necessary to study the reason for water migration to the freezing zone.

From the discussion so far, you should be able to visualise a 'free energy gradient' for water. This gradient declines from the high free energy possessed by water well away from the frozen zone to the low free energy of water adjacent to ice. Now, a transfer of mass tends to occur along a free energy gradient, and so there is a movement of water to a freezing surface. This movement will be affected by rock permeability, temperature and water availability. The gradient can be expressed as a suction force on water, which may reach $10^6\,\mathrm{N\,m^{-2}}$. Of course, there is a similar pressure exerted on rock material.

The presence of openings, particularly cracks (*see* Fig. 6.2), in virtually all rocks may be exploited by the pressure, which is called the *stress intensity factor*, K_I. It is affected by crack geometry and ice pressure. It is resisted by the tensile strength of the rock, or *fracture toughness*, K_C. Clearly, when $K_I > K_C$, cracks grow and mechanical weathering occurs. Crack growth ceases when K_I falls below a critical value K_*, the *stress corrosion limit*. For

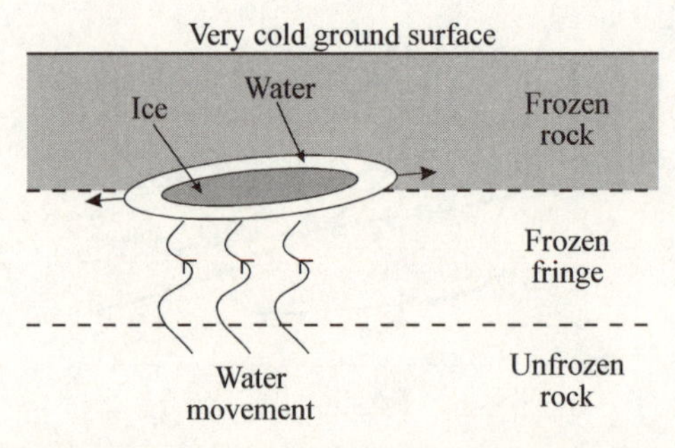

FIGURE 6.2 The freezing of a cracked rock. The likely direction of crack extension is shown by the arrows (⟶). (*Source*: based on Walder, J. and Hallett, B. 1985: A theoretical model of the fracture of rock during freezing. *Geological Society of America Bulletin* **96**, 336–46.)

rocks, estimates suggest that

$$\frac{K_*}{K_C} \sim 0.3\text{–}0.4$$

This model, then, uses the dynamic behaviour of water at sub-zero temperatures to bring about crack growth within rock.

Experimental and theoretical support

The 'ice growth' hypothesis is supported by experimental and theoretical findings, which are also consistent with each other.

- **Experimental results** show that:
  - Water and ice may coexist at sub-zero temperatures and water may migrate to freezing centres.
  - Cracks in rock propagate distances of about a grain diameter (~0.1–10 mm) in between 1 and 100 days. This gives average crack-growth rates of about 10^{-11} to 10^{-7} m s^{-1}.
- **Numerical calculations** have been used to calculate crack growth rates for a granite and a marble under varying conditions and making certain assumptions. The calculated crack-growth rates are consistent with those found by experiment.
- **Physical principles** readily account for the results of numerical calculations which predict that crack growth rates should be highest in the temperature range –4 to –15°C, and diminish when temperatures are higher and lower.

When the frozen fringe is relatively permeable, the maximum possible pressure is reached in only a few hours. This pressure is, however, only a few MPa and so crack growth rates are very low. At very low temperatures the rate of flow of water through the frozen fringe is extremely low and it could take years to reach maximum ice pressure. At intermediate temperatures cracks grow rapidly because the frozen fringe is still quite permeable, and the 'stress intensity factor' can reach significant values quite quickly.

This model ties together a large amount of data obtained through laboratory and field studies. More specifically it links:

- **rock properties**, including:
  - *pore size*, which affects freezing temperature;
  - *permeability*, which controls water movement at sub-zero temperatures;
  - *fracture geometry and toughness;*
- **temperature regime**, which most favours crack propagation when the temperature ranges from –5 to –15°C, and when cooling rates are quite low, less than 0.1–0.5°C h^{-1};
- **moisture availability**, which contributes most effectively when the water content is high and where supply is ample.

The model is consistent with physical principles. It avoids the need for freeze–thaw cycles and for rapid freezing rates. It offers a good working hypothesis to explain the mechanics of frost weathering.

6.1.3 Hydration shattering

Theory

This idea is based on the view that frost action is not the only important cold-climate process and that the behaviour of unfrozen water at very low temperatures may have important weathering consequences.

The idea is that individual polar water molecules will become oriented to charged mineral surfaces in such a way that one end of a molecule bonds to the mineral surface while the other end projects away (*see* also section 6.3). These molecules may not readily freeze at sub-zero temperatures as there may be insufficient energy for them to be reoriented. An alignment of molecules is thus developed, and their rigidity increases as the temperature falls.

Repulsion forces may be set up in very narrow cracks or pores less than 5 μm in diameter when the free ends of one set of molecules approach the similarly charged ends of the set of molecules on the other side of the crack (or pore). The magnitude of the repulsion increases as the temperature falls, and is helped by the expansion of water that takes place when the temperature falls below 4°C. Ultimately, rock breakdown may occur.

Contribution of laboratory studies

Investigations of the breakdown of clay-rich dolomites and shales have tended to support the theory. In rock samples that responded readily to 'frost' weathering, typically more than 50 per cent of their contained water was still unfrozen, even at temperatures below −40°C. Such unfrozen water was believed to be in an 'ordered' state. Detailed examination showed that thin veneers of clay, with residual negative charges, surrounded magnesium carbonate crystals (in dolomite) in the samples and that the unfrozen water molecules were adsorbed to the clay surfaces.

Work on schists has also supported the hydration process at temperatures around freezing. The relative significance of frost action and hydration weathering was, however, hard to determine, but it was estimated that up to about 25 per cent of the debris released was due to hydration effects.

Support for the theory is rather limited, being based so far on studies of carbonates and schists. There is a risk, therefore, in making extrapolations to other rock types, although the mechanism has been proposed for silicate-rich rocks. It would be reasonable to suggest that clayey rocks generally may be susceptible to the process.

6.1.4 Ice crystal growth

Theory

Investigations have suggested that there may be a parallel between the growth of, for example, salt crystals from a solute and the development of crystals from freezing water. In the former case the mechanism involves the deposition of solutes to a growing crystal in such a way as to exert pressure on adjacent confining surfaces. This pressure is at its maximum in the direction of greatest resistance. A similar situation is seen for freezing water. Certain conditions would have to be met for the process to work effectively. These include a slow rate of freezing, allowing the movement of supercooled water to zones of crystallisation, and a permeable texture which would encourage the migration of water.

Contribution of laboratory studies

The pressure exerted by the growth of polycrystalline ice from a melt has been measured in the laboratory and found to be about 20 kPa.

Measured stresses are well below the tensile strength of rocks. However, the process may exploit existing planes of weakness, which may fail further at stresses well below the **bulk** tensile strength of the material. The process is consistent with the water migration theory (*see* section 6.1.2) and may throw light on details of that process.

6.1.5 Hydraulic pressure

Theory

This idea comes from engineering studies of the reaction of concrete to freezing. It is based on the view that high and destructive water-based pressures may develop in front of a freezing plane.

You can understand the process by setting up a 'thought experiment'. This means that you work out ideas and their consequences in your head. Imagine a block of material that has been in contact with water before freezing begins. Assume that the contact surface and immediate interior of the block will be saturated. Farther in, the water content will be less. Now imagine the external temperature falling below the freezing point. The surface of the block will freeze, forming a barrier to water movement. Farther in, the near-surface water in coarser pores will also freeze, and the associated expansion will push any unfrozen water towards the interior. In the case of a fine-grained porous material like concrete, there will be frictional resistance to this water movement. If the resistance is high enough the material will be broken up.

The magnitude of the hydraulic pressure developed depends on the rate of movement of the water, which in turn is controlled by the freezing rate. The rate of freezing at a point in a material is determined by the distance of that point from the surface. The relationship is such that the greater the distance of the point from the surface, the lower is the freezing rate. This suggests that the effectiveness of the process may be influenced by the size of the 'weathering' object.

Engineers studying the behaviour of concrete during freezing have found this process convincing. You might think it reasonable that it should also disrupt natural materials which may have characteristics of porosity and permeability which are similar to those of concrete.

6.2 Salt weathering

Salts, which are chemical compounds formed from reactions between acids and bases (*see* Box 6.1), can cause rock breakdown. This is because of the pressures they exert, either when they crystallise from solution or when, in the crystal form, they expand on heating or hydration.

Many salts are water soluble, so they are only widely found in arid and semi-arid areas such as hot and cold deserts (*see* Box 6.2). The limited supplies of water mean that salts are not removed in solution from the environment. There may, however, be sufficient water for some salts to be hydrated (*see* Box 6.2). Salts are also important in urban environments, as shown in Box 6.1.

For salts to be effective weathering agents, they need to enter rock pores, typically in solution. The relevant types of porosity and some useful measures of water content are summarised below (you should remember most of these ideas from Chapter 3):

- **porosity** is the total volume of pore space expressed as a percentage of the bulk volume of the rock;
- **microporosity** is the proportion of micropores (very small pores) expressed as a percentage of total available pore space;
- **water absorption capacity** is a measure of the amount of water absorbed in a specified time;
- **saturation coefficient** is the amount of water absorbed in 24 h when a sample is totally immersed. It is expressed as a 'fraction' of the volume of available pore space.

Some investigators have suggested that the ability of solutions to penetrate rock also varies with the salt involved. For example, there is some evidence that a sodium chloride (NaCl) solution is absorbed more rapidly and achieves a greater degree of saturation than a solution of sodium sulphate (Na_2SO_4).

Salts are believed to cause rock breakdown by exerting three types of expansionary force, due to thermal expansion, hydration and crystallisation.

6.2.1 Thermal expansion

This idea is very similar to that described in section 6.4 on insolation weathering, where the temperature ranges recorded in hot deserts are described.

Box 6.1 The formation of a salt

A salt is an ionic solid (*see* Chapter 1). It is produced by the reaction of an acid and a base, with water as a by-product:

$$\text{acid} + \text{base} \longrightarrow \text{salt} + \text{water}$$

A simple laboratory example might be

HCl	+	NaOH	$\longrightarrow$	NaCl	+	H_2O
hydrochloric acid	+	sodium hydroxide	$\longrightarrow$	sodium chloride	+	water

An example of salt formation which has practical importance may be observed on buildings made of limestone (e.g. churches) in polluted areas. Such salt formation involves several stages.

1. Sulphur dioxide, perhaps derived from the burning of fossil fuels, combines with rainwater to form sulphurous acid:

$$SO_2 + H_2O \longrightarrow H_2SO_3$$

2. This is rapidly oxidised to form sulphuric acid:

$$H_2SO_3 + O \longrightarrow H_2SO_4$$

3. The weakly acid rainwater (remember that there are several other sources of acidity, including dissolved carbon dioxide) reacts with the limestone to produce salt:

H_2SO_4	+	$CaCO_3$	$\longrightarrow$	$CaSO_4$	+	H_2O	+	CO_2
sulphuric acid	+	calcium carbonate	$\longrightarrow$	calcium sulphate (salt)	+	water	+	carbon dioxide

The salt is initially held in solution. Part may then be lost in run-off and part may be washed into rock pores where the salt weathering processes take place.

For weathering to occur, the thermal expansion coefficient of a salt should exceed that of the surrounding rock. For a salt such as sodium chloride (NaCl), a temperature rise of about 50°C produces a volumetric expansion of 1 per cent, which is greater than that of any enclosing rocks. This sets up a tensile stress, to which rocks have least resistance. The process may be particularly effective in granitic rocks which may already be under internal stresses.

However, while the process is theoretically reasonable, it has not yet been demonstrated by laboratory experiment and is less important than hydration and crystallisation pressures.

Box 6.2 Common salts and their hydrates in hot and cold deserts

- Some common salts in deserts include calcium carbonate as well as the *sulphates* (containing the SO_4^{2-} group), *nitrates* (containing the NO_3^- group) and *chlorides* (containing Cl^-) of sodium and magnesium. Others of course occur, but these seem to have most weathering significance. In hot deserts gypsum (see below), sodium chloride (NaCl), sodium sulphate (Na_2SO_4) and calcium carbonate ($CaCO_3$) appear to be most widespread. In cold deserts, sulphates of sodium and magnesium are two of the most common salts.
- Water molecules may associate with a salt. The resulting compound is called a *hydrate* (*see* section 7.7). A range of hydrated water molecules is found, for example:
  - $Na_2SO_4 \cdot 10H_2O$ is *mirabilite*, a *deca*hydrate (the prefix shows the number of hydrated water molecules);
  - $MgSO_4 \cdot 7H_2O$ is *epsomite*, a *hepta*hydrate;
  - $MgSO_4 \cdot 6H_2O$ is a *hexa*hydrate;
  - $CaSO_4 \cdot 2H_2O$ is *gypsum*, a *di*hydrate;
  - $MgSO_4 \cdot H_2O$ is *kieserite*, a *mono*hydrate.

  You should have spotted that for the same salt, $MgSO_4$, different hydrates are possible. Another example is the salt $CaSO_4$, which may have the *hemi*hydrate form $CaSO_4 \cdot \frac{1}{2}H_2O$. Compare this with gypsum (above).
- A salt lacking water is described as 'anhydrous', e.g. the sodium sulphate *thenardite*, Na_2SO_4. *Anhydrite* refers to anhydrous calcium sulphate, $CaSO_4$.

6.2.2 Hydration pressure

When salts hydrate (*see* Box 6.2) there may be a considerable increase in volume. For example, when the monohydrate $MgSO_4 \cdot H_2O$ hydrates further to the heptahydrate $MgSO_4 \cdot 7H_2O$, the volume expansion exceeds 170 per cent, and when thenardite (Na_2SO_4) hydrates to mirabilite ($Na_2SO_4 \cdot 10H_2O$) the increase in volume is greater than 300 per cent.

Generally, the greater the degree of hydration, the greater the pressure generated. Hydration pressures of up to 63 bar have been recorded in the laboratory, while in theory higher pressures are possible and compare with rock tensile strengths of between 20 and 200 bar.

The process gains further support from the observation that both the sulphates referred to may pass through a complete hydration–dehydration cycle in less than 24 h. However, for this to happen the temperature must reach the transition point (*see* Box 6.3) at which the anhydrous (or less hydrated) salt becomes fully hydrated. The temperature must then fall for the process to be reversed.

Box 6.3 The transition point

This is the temperature at which one crystalline form of a substance, such as an anhydrous, or perhaps less hydrated salt, is converted into another, such as a more hydrated salt. The process is reversible. For example, the transition point for the change

$$Na_2SO_4 + 10H_2O \rightleftharpoons Na_2SO_4 \cdot 10H_2O$$

is 32.4°C. For the change

$$MgSO_4 \cdot 6H_2O + H_2O \rightleftharpoons MgSO_4 \cdot 7H_2O$$

it is 48.2°C.

6.2.3 Crystallisation pressure

Crystallisation can be brought about in two ways:

- by increasing the concentration of a solution at a constant temperature, usually by evaporation;
- by lowering the temperature of a solution that is close to being saturated.

We will now develop these two ideas. You will find it helpful to refer to Fig. 6.3 where the ideas are shown graphically.

When a salt solution in a rock void starts to evaporate, perhaps due to high temperatures, the concentration of the salt increases. A stage is reached

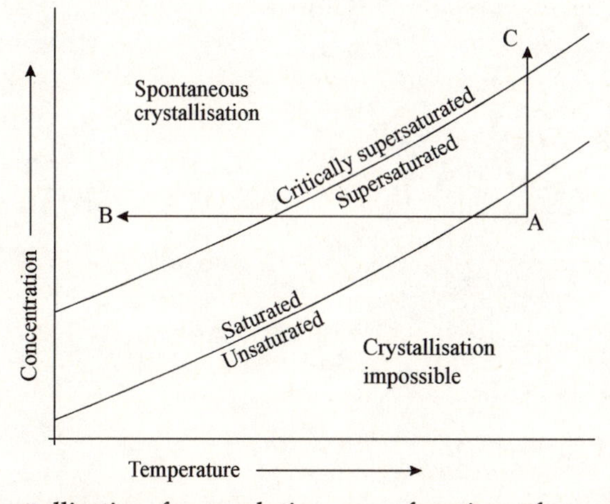

FIGURE 6.3 Crystallisation from solution as a function of concentration and temperature. Supersaturation can be reached in two ways: A ⟶ B by a fall in temperature at constant concentration, and A ⟶ C by an increase in concentration by evaporation at constant temperature. (Based on Winkler, E.M. and Singer, P.C. 1972: Crystallization pressure of salts in stone and concrete. *Geological Society of America Bulletin* **83**, 3509–14.)

where the salt is just maintained in solution. The solution is then said to be 'saturated'. Up to this point crystallisation is not possible. A further reduction of the water content can have two consequences. The salt may crystallise, but this is a very slow process and has to take place round foreign particles or 'nuclei' (e.g. dust). Alternatively solution is maintained, in which case it is said to be 'supersaturated'. Crystallisation is postponed, but when it occurs it is very rapid.

If a solution is cooled, the effect will depend on the concentration. If the solution is far from saturation, then no crystallisation will take place. If it is near to saturation, then crystallisation may take place. Supersaturation can, however, occur, as in the situation discussed above.

The growth of crystals gives rise to crystallisation pressure. This is usually well in excess of the tensile strength of rock. For example, the crystallisation pressure of halite (NaCl) is about 650 bars, when it is oversaturated by a factor of two. Study Table 6.1 which shows some of the pressures that can be generated. Their nature and magnitude depend on the following factors.

- **The degree of supersaturation, if any, of a saline (i.e. salty) solution.** Generally, crystallisation pressure is proportional to the degree of supersaturation.
- **Crystallographic properties.** These include crystal-forming potential, crystal habit (form), the rate of crystal growth, the crystal size achieved before fracture, and any change of crystal style with alternate hydration and dehydration.

TABLE 6.1 Calculated crystallisation pressures for some salts

Salt	Formula	Density ($g\,cm^{-3}$)	Crystallisation pressure (atm)			
			$C/C_S = 2$		$C/C_S = 10$	
			0°C	50°C	0°C	50°C
Epsomite	$MgSO_4 \cdot 7H_2O$	1.68	105	125	350	415
Gypsum	$CaSO_4 \cdot 2H_2O$	2.32	282	334	938	1110
Halite	NaCl	2.17	554	654	1845	2190
Mirabilite	$Na_2SO_4 \cdot 10H_2O$	1.46	72	83	234	277
Natron	$Na_2CO_3 \cdot 10H_2O$	1.44	78	92	259	308
Hexahydrate	$MgSO_4 \cdot 6H_2O$	1.75	118	141	395	469

Note: the calculations are based on the Riecke principle:

$$P = \frac{RT}{V_S} \cdot \ln \frac{C}{C_S}$$

where P is the pressure exerted by the growing crystal, in atmospheres (1 atm = 1.01325 bar = 1.01325×10^5 Pa); R is the gas constant (0.0821 atm mol^{-1} K^{-1}); T is the absolute temperature (K); V_S is the molar volume of solid salt (l mol^{-1}); C/C_S is the supersaturation ratio, where C is the existing solute concentration (supersaturation) and C_S is the saturation concentration, both measured in mol cm^{-3}; and ln (natural log) = 2.303 $\log_{10}$.

Source: Based on Winkler, E.M. and Singer, P.C. 1972: Crystallisation pressure of salts in stone and concrete. *Geological Society of America Bulletin* **83**, 3509–14.

Salt weathering has been studied under controlled laboratory conditions. A common method is to expose different rock types to a range of salts in aqueous solution under a variety of simulated climatic conditions. A 'climatic cabinet' is often used which allows environmental factors such as temperature and humidity to be artificially controlled.

Climatic sequences can be accelerated in order to speed up the processes and produce results rapidly. Such laboratory approaches allow the experimenter to test hypotheses about salt weathering and to monitor processes with a precision that cannot be achieved in field investigations.

However, you should note three weaknesses in the laboratory approach.

- *The problem of reproducing the complexity of natural climatic conditions.* For example, the effect of insolation (exposure to the sun) cannot be included in the programming of a 'climatic cabinet', and it is also almost impossible to duplicate the natural heat and humidity changes that take place just above the ground surface. In any event, these are often not known for the natural environment. A related problem concerns the simulation of climatic cycles. The common use in experiments of extreme cycles of temperature and humidity does not match natural events.
- *The extent to which experimental rock samples are representative of natural material.* Test fragments may not have the size and shape characteristics of those in the field. It is also difficult to reproduce natural stress conditions in the laboratory.
- A final difficulty involves *the nature of the solutions used in laboratory experiments.* These typically involve a single salt–hydrate system (e.g. $Na_2SO_4 \cdot 10H_2O$) while complex mixtures are most common in nature. Thirty salts have been recorded in cold regions.

6.2.4 Influence of salts on frost weathering

Salts are a significant surface and near-surface feature of cold environments, as the lack of water means that they are not readily lost in solution. Observation shows that three salts are particularly important in such regions: sodium chloride (halite), sodium sulphate (thenardite and mirabilite) and magnesium sulphate (epsomite and hexahydrate). You might think that the presence of such salts may have an effect on frost weathering and this problem has received much laboratory investigation.

The results of such studies have not, however, been very consistent. Most workers have found that the presence of salts increases frost weathering under experimental conditions, but some have observed that salts may have little effect on, or even inhibit, frost action. It is not easy to reconcile such inconsistent results because of the differing experimental procedures and variety of rock type employed.

Recent work suggests that the influence of salts on frost weathering is complex and varies with the type and concentration of salt, the intensity of the freeze–thaw cycles and the rate of freezing and thawing. As an example, it appears that the greatest rate of breakdown is experienced by rock samples in dilute (5.5 per cent) sodium chloride solutions subject to intense (to –30°C) freeze–thaw regimes. Rock samples show less deterioration when placed in concentrated solutions (12.5 per cent) of the same salt. On the other hand, samples in concentrated sodium chloride solutions subjected to mild (to –10°C) freeze–thaw regimes show little or no destruction. Sodium sulphate, either on its own or mixed with sodium chloride, encouraged rock breakdown under both freezing regimes. By contrast, the breakdown of rock samples was actually discouraged when they were placed in a solution of magnesium sulphate and subjected to intense freezing.

These experimental findings can be partly explained through a consideration of the phase changes (i.e. changes of state) that occur in, for example, a sodium chloride solution as the temperature falls. At –10°C (mild freeze–thaw regime) a mixture of ice and solution remains, partly because the presence of salts lowers the freezing point, and partly because the release of latent heat at crystallisation slows down ice formation. Consequently the weathering effect is minimal. At –30°C (intense freeze–thaw regime) all of the liquid is crystallised to form a *cryohydrate*, a mixture of water ice and salt crystals. Crystallisation occurs because the temperature is below the eutectic temperature (*see* Box 6.4) at which crystallisation of the whole solution occurs. Rock stresses are high and breakdown is probable. A finding by some experimenters that a solution of sodium chloride actually retarded frost weathering may be explained by a possibly incomplete freezing of the rock sample at the temperature used.

The idea of phase changes does not, however, provide a complete explanation of breakdown when rocks are frozen in salt solutions. It seems that variation in the shape, size and strength of salt and ice crystals must also be considered.

These general findings are perhaps a little artificial, in that the rock samples were totally immersed in the solutions. A more realistic circumstance

Box 6.4 Eutectic temperature and concentration

The *eutectic temperature* is the temperature at which a solution crystallises. The solid crystallising out has the same concentration as the solution from which it comes. Different salt solutions have different eutectic temperatures; e.g. –21°C for sodium chloride and –1.2°C for sodium sulphate. The *eutectic concentration* is the concentration at which crystallisation occurs. If a concentration is less than the eutectic concentration, cooling to the eutectic temperature produces only ice crystallisation. If the concentration is greater than the eutectic concentration, crystallisation of the salt will occur above the eutectic temperature.

occurs when freezing takes place under non-immersed conditions, as might be the case in, for example, a scree slope. Recent experimental work confirms that the presence of a salt, especially sodium sulphate, increases rock weathering at temperatures below 0°C, but at a lower intensity than with fully immersed samples. It appears that the dominant process is salt crystallisation rather than ice formation.

6.3 WEATHERING BY WETTING AND DRYING

In many natural environments, exposed rock surfaces experience cycles of wetting and drying when, for example, short rain events are followed by periods of evaporation. Such changes may bring about rock breakdown. The process is closely related to hydration shattering (*see* section 6.1.3). In this section we shall study the process, and also consider its significance for frost-related weathering and salt weathering.

You can understand the process by following the events believed to take place at the surface of a fractured rock or mineral, which will typically show unsatisfied electrostatic bonds. If such a surface occurs inside a rock unit, perhaps as a wall of a fine crack, and is wetted, then polar water molecules will be attracted to it and in due course will make up a layer of adsorbed water. The word 'a*d*sorbed' means 'on the surface'; 'a*b*sorbed' means 'taken internally' – like a sponge a*b*sorbs water. The addition of further water may bring about a swelling pressure within such a crack, which may create strain. If now water is lost, perhaps by evaporation, only the most tightly held water molecules will remain. The sides of the crack may now be pulled together as attractive forces may occur between residual water molecules on opposing faces of the crack (*see* section 1.4). Where particles are involved, similar principles may apply. Cycles of wetting and drying will tend to give rise to expansion and contraction, with cracking and flaking of rock.

Shale (a compacted mud, which splits easily into thin layers) is particularly susceptible to this process. Other rock types, e.g. sandstone, limestone, granite, basalt and schist (an altered rock having flaky minerals aligned to form a cleavage) are also affected. A study of vulnerable rock types suggests that certain rock properties favour weathering by wetting and drying. These include

- the presence of clay minerals, which encourage water adsorption;
- structural weaknesses, such as cleavage planes and planes of schistosity, which encourage water entry;
- low tensile strength;
- pore size and distribution.

The importance of structural and pore-size factors has been shown by recent experimental work. This has drawn attention to some implications of the wetting and drying process. The water-holding property (i.e. percentage saturation) of sandstone and dolerite (a dark-coloured, medium-grained igneous rock), has been shown to increase over time. This can be explained by an increase in the size and/or number of pores and microfissures as wetting and drying proceeds.

This change has important implications for both frost and salt weathering, and especially for their experimental study. It seems that part of the breakdown attributed to these processes may in fact be due to wetting and drying, and that the rock-weakening effect of the latter mechanism may accelerate the action of the other processes. Wetting and drying is an important mechanism in its own right, and may increase the effectiveness of the other processes.

6.4 Insolation weathering

In this section you will study the contribution that *temperature change* may make to the physical breakdown of exposed rocks. At first sight the process appears quite straightforward. A surface layer of rock expands when it is heated by the sun's rays ('insolation' is exposure to the sun), and then contracts when the heat source is temporarily cut off. Breakage occurs when the stresses due to expansion and contraction exceed the rock's elastic limit (*see* Chapter 3). Possible field evidence for the process can be seen in many hot, rocky deserts, where spreads of apparently shattered stone fragments that seem to be chemically unaltered have been described. This weathering process has been called *thermal fatigue*, *insolation-related weathering* and *thermoclastis* ('breaking by heat'). *Insolation weathering* is perhaps the simplest description.

The process is surprisingly complicated and many of the details are not well understood. A helpful way of organising your study of the topic is to divide it into four sections: the nature and causes of temperature change, the influence of rock properties, the contribution of experimental studies, and a summary of the modifications needed to the simple model described above.

6.4.1 The nature and causes of temperature change

Rock surfaces are exposed to two main sources of heat: the sun, through solar radiation, and natural fire. Either source may raise the surface temperature, which then falls when the heat source is cut off. This behaviour introduces the important ideas of *temperature range* and *rate of temperature change*. The temperature range is the difference between the highest and lowest temperatures over a specified period of time, usually 24 h, while the rate of temperature change is the amount by which temperature changes during a specified period of time. Look at Box 6.5 for an illustration of these ideas.

Box 6.5 Temperature range and rate of temperature change

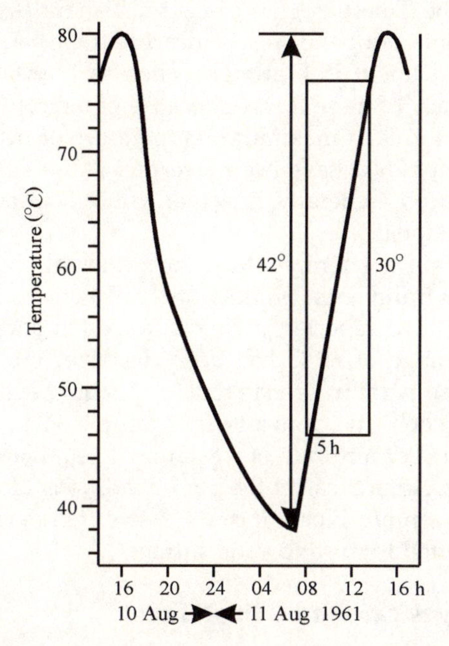

The **temperature range** is the difference between minimum and maximum temperatures over a time period. In the figure, for a basalt surface at Wour, Tibesti, Sahara, it is approximately 42°C over 24 h. The **rate of temperature change** is the amount of change in unit time. In the graph between 0800 and 1300 h it is approximately 6°C h^{-1}, i.e.

$$\left(\frac{76^\circ - 46^\circ}{5\,\text{h}}\right)$$

The data should be seen as illustrative.

(*Source of data*: Peel, R.F. 1974: Insolation weathering: some measurements of diurnal temperature changes in exposed rocks in the Tibesti region, central Sahara. *Zeitschrift für Geomorphologie* Suppl., **21** 19–28.)

Temperature changes associated with solar heating and subsequent cooling

The nature of solar heating is affected by a number of external factors, including latitude, altitude, aspect (orientation of the rock surface), time of year, nature of the cloud cover, wind speed and air temperature. Most studies of solar heating have been carried out in hot or cold deserts where a high proportion of the land surface may be bare rock.

Very high surface temperatures and ranges have been recorded in **hot deserts**. Studies in the Tibesti region of central Sahara (*see* Box 6.5) have shown surface daytime temperatures of nearly 80°C (perhaps enough to cook an egg), falling to about 35°C at night. There is, however, some uncertainty over the accuracy of these figures because of instrumental problems. In Death Valley, California, a maximum temperature of nearly 73°C and a minimum of just under 28°C have been recorded (August 1992) for a rock surface over a 24 h period. Generally, however, ground surface temperatures above 60°C are exceptional.

Surprisingly high surface temperatures and rates of change have been recorded in **cold environments**. Studies on Ellesmere Island (Northwest Territories, Canada) have reported a rock surface temperature of 39.7°C, and a temperature range of 60°C has been described for Victoria Land, Antarctica. Low air temperatures mean that very rapid rates of temperature change may occur in such environments, for example when a passing cloud obscures the sun. A maximum rate of $0.8°C\,min^{-1}$ (equivalent to $48°C\,h^{-1}$) has been recorded in Antarctica. In the cold conditions of the Karakorum mountain range of Kashmir, North India, a rate of temperature change of $2.7°C\,h^{-1}$ has been noted for a sandstone surface.

Temperature changes caused by bush fires

Bush fires are infrequent events, but when they occur they raise rock surface temperatures to high values over short time periods. The highest recorded temperatures have exceeded 800°C, and values of 500°C are quite common. The time taken for such temperatures to be reached is quite short, and may be less than 10 min. The term *thermal shock* describes the impact of such a rapid change. The thermal shock experienced by rocks exposed to such conditions is likely to have destructive effects.

6.4.2 Influence of rock properties

A knowledge of the surface temperature does not tell us about temperatures inside the rock. To understand this you need to study a number of physical properties and how they affect the reaction of a rock to surface heating or cooling. Some important properties are as follows.

Albedo

Some of the solar radiation falling on a surface is reflected. The amount reflected is the *albedo* of the surface and is expressed as a percentage of total solar radiation. It is important because it is a measure of a rock's ability to absorb heat and so raise its temperature. A basalt (an igneous rock poor in silica) may have an albedo of about 12 per cent, granite about 18 per cent, chalk about 25 per cent and desert sand about 40 per cent.

Table 6.2 Some examples of thermal conductivity

Rock type	Thermal conductivity ($W\,m^{-1}\,K^{-1}$)
Basalt	0.96
Portland limestone	1.53
Granite	1.65
Chalk	1.72

Source: Warke, W.A. and Smith, B.J. 1994: Short-term rock temperature fluctuations under simulated hot desert conditions: some preliminary data. In Robinson, D.A. and Williams, R.B.G. (eds) *Rock weathering and landform evolution*. Chichester: John Wiley, 57–70.

Thermal conductivity

When the temperature of a surface is raised, heat is normally conducted to the subsurface. *Thermal conductivity* is a measure of the rate at which heat is transmitted through a substance and is recorded in $W\,m^{-1}\,K^{-1}$ (watts per metre per kelvin – note that a change of 1 K is the same as 1°C (Celsius) but absolute degrees (K) are used in the official international system of units). Metals generally have a high conductivity, as you will realise if you touch a metal spoon left in a hot liquid. The figure for aluminium is about 240. Rocks, on the other hand, have a low thermal conductivity, and some figures are given in Table 6.2. An important consequence of a low thermal conductivity is that a marked temperature gradient may result, and some field measurements are given in Table 6.3.

In more detailed work it was shown that the temperature gradient may be much steeper in the outer few millimetres of a rock. The existence of a temperature gradient means that fluctuations in surface temperature will not affect the temperature at a certain depth. For example, it was found that the sand dunes of Tibesti had daily temperature ranges of 38–39°C at the surface which fall to 1–2°C at 30 cm depth, and were not observed at 75 cm.

You should appreciate that a steep temperature gradient is likely to be associated with a similarly steep stress gradient in the upper layers because of changes in the amount of expansion.

Table 6.3 Examples of recorded rates of temperature decline from the surface to the interior of certain rocks

Rock type	Location	Temperature fall	Temperature gradient ($°C\,cm^{-1}$)
Quartz monzonite[a]	South California	15°C in 30 cm	0.5
Sandstone	Tibesti, Sahara		0.85
Limestone		10.1°C in 10 cm	1
Desert sand	Tibesti		1.2

[a] A coarse-grained igneous rock.
Source: Various.

Box 6.6 The coefficient of linear thermal expansion (α)

When a rod of material of length L_O is heated, it undergoes a temperature change of ΔT and an increase in length of ΔL. The *fractional change* in the length of the rod is

$$\frac{\Delta L}{L_O}$$

Now

$$\alpha = \frac{\text{fractional change in length}}{\text{temperature change}} = \frac{\Delta L / L_O}{\Delta T}$$

i.e.

$$\Delta L = \alpha L_O \Delta T$$

The units of α are reciprocal degrees, either 1/°C or 1/K. Note that the linear expansion coefficient for an isotropic (*see* Chapter 3) solid is about one-third its volume expansion coefficient.

Now let us put some figures into the equations. Consider the case of a rod, $L_O = 10$ cm, cut from a quartz crystal ($\alpha = 9 \times 10^{-6}\,°C^{-1}$) and heated through 10°C:

$$\Delta L = (9 \times 10^{-6}) \times 10 \times 10$$
$$= 0.0009\,\text{cm} \quad \text{or} \quad 9 \times 10^{-4}\,\text{cm}$$

Coefficient of thermal expansion

When a solid is heated it expands, and when cooled it contracts. Solids vary in their response to temperature change, and an index of the degree to which they respond to an increase in temperature is their *coefficient of thermal expansion*. For weathering studies a commonly used index is the *coefficient of linear thermal expansion* (*see* Box 6.6, where you can also see that the actual changes in length are quite small).

The position is quite complicated as far as rocks are concerned, particularly at the mineral (crystal) scale. Individual crystals have varying coefficients, and even within a single crystal coefficients may vary with the direction of measurements (*see* Table 6.4).

Expansion may generate very high stresses, as you will see from the case of quartz. When heated this mineral can generate an expansionary force roughly equal to its compressive strength (*see* Chapter 3). This is about 25 000 kg cm^{-2} parallel to, and 22 800 kg cm^{-2} normal to, the *c*-axis (see note to Table 6.4).

6.4.3 The contribution from experimental studies

By now you will have gained some appreciation of the number of factors involved in insolation weathering. You will not be surprised to read that

TABLE 6.4 Some coefficients of linear thermal expansion for minerals, showing variation with direction

Mineral	Coefficient of linear thermal expansion ($\times 10^{-6}\,K^{-1}$)				
	Parallel to *a*-axis	Parallel to *b*-axis	Parallel to c-axis	Normal to c-axis	(Same in all directions)
Rock salt, NaCl					40
Fluorite, CaF_2					9
Calcite, $CaCO_3$			26	6	
Quartz, SiO_2			9	14	
Aragonite, $CaCO_3$	10	16	33		

Note: The *a*-, *b*- and *c*-axes are imaginary reference lines (crystallographic axes) that run parallel to the edges of the unit cell.

many investigators have concentrated on experimental studies, which allow considerable control of many of the factors. An important approach has been to focus on simulation studies, where the assumed natural processes are repeated in the laboratory. Three sets of field conditions have been copied.

Where daily temperature change appears important

Much of the early experimental work was an attempted simulation of daily temperature changes on the assumption that these caused rock breakdown. In an early investigation, Griggs (1936) experimented on coarse-grained granite, which might have been expected to show the effect of varying coefficients of expansion among the constituent minerals. He heated and cooled his samples through a temperature range of 110°C for a period equivalent to 240 years of daily change of that magnitude. In the absence of water, no breakdown was observed. In later work, Goudie (1974) experimented on a silica-cemented sandstone and on chalk, subjecting his samples to 58 and 43 temperature cycles respectively. Each cycle, consisting of 1 h at 17–20°C, 6 h at 60°C and 17 h at 30°C, broadly corresponded to the daily temperature range for the ground surface in a hot desert such as the central Sahara. No rock breakdown was reported. Goudie also experimented on the possible effect of the thermal expansion of the salt sodium nitrate, which was thought to be potentially destructive due to its relatively high coefficient of linear thermal expansion. Rock samples were saturated with a solution of sodium nitrate and then subjected to the temperature cycles described above. Again, no breakdown was seen.

Where short-term fluctuations of rock temperature occur

You have seen that, especially in cold environments, rapid temperature fluctuations may occur. Hall and Hall (1991) attempted to simulate such conditions by exposing rock samples, in an ambient temperature of about −19°C, to variable warming cycles by means of infrared lamps. During the

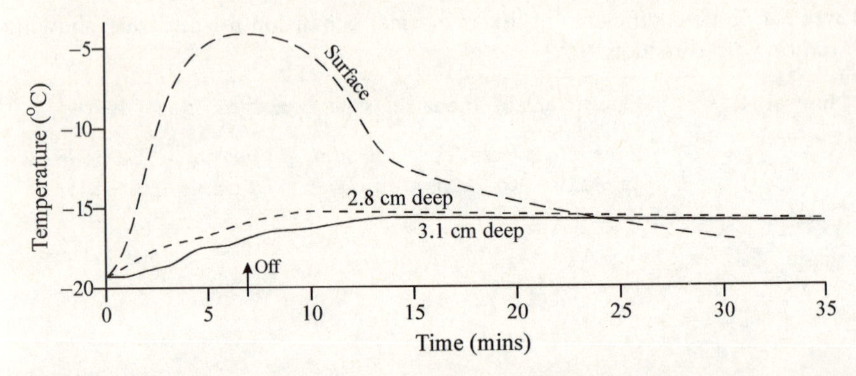

FIGURE 6.4 Temperature data from a block of rock showing how internal temperatures continue to rise after the removal of the heat source brings about surface cooling. (*Source*: Hall, K. and Hall, A. 1991: Thermal gradients and rock weathering at low temperatures: some simulation data. *Permafrost and Periglacial Processes* **2**, 103–12.)

experiments the temperatures of the air, of the rock surface and at depths of 2.8 and 3.1 (and 3.3) cm were recorded. The investigators found that:

- the rate of temperature change was greatest at the rock surface and decreased with depth;
- the rate of temperature change was extremely high for short periods of time, with values greater than 8°C min^{-1} being recorded for the surface; it appeared that rates of 2°C min^{-1} operated to a depth of 2.2 cm;
- the interior of the rock sample continued to warm when the surface was cooling following removal of the heat source (*see* Fig. 6.4).

Short-term temperature fluctuations under 'hot desert' conditions have also been simulated, and the results show the importance of *albedo* and *thermal conductivity*. Under an environmental temperature of 40°C, the surface of a basalt rose to a maximum of 53°C as a consequence of its low albedo. At 2.5 cm depth the temperature was 48.9°C (i.e. 4.1°C lower). This steep gradient reflected low thermal conductivity. For chalk, with a high albedo, the maximum surface temperature was 37.2°C, falling by only 1°C at 2.5 cm depth as a consequence of its high conductivity.

Where natural fires occur

Quite dramatic findings have emerged from attempts to simulate the extreme conditions of bush fires, when ground surface temperatures may exceed 500°C, and very high temperatures can be maintained for some minutes.

Experimental work by Goudie *et al.* (1992) focused on the behaviour of a number of rock types which were subjected to very high temperatures. They studied the response of seven rock types in two experimental situations. The first involved heating samples for 5 min at one of a number of temperatures

TABLE 6.5 Change in modulus of elasticity for a range of rock types subject to differing experimental conditions simulating the effect of bush fires

Rock type	Percentage change (decrease) in modulus value (%)	
	a	b
White granite	61.8	37.0
Shap granite	52.2	17.8
Portland stone	42.2	14.4
White marble	38.0	16.0
Gabbro	17.5	2.0
York stone	2.7	3.9
Slate	0.0	0.0

a. After 5 min heating, then cooling, at a starting temperature of 500°C. Cycle repeated five times at intervals of 24 h.
b. After heating at 500°C for 1 min.
Source: Goudie, A.S. *et al.*, 1992.

between 50 and 900°C, and then allowing them to cool. Five such cycles were carried out, at intervals of 24 h. The second experiment involved samples being heated at a constant temperature of 500°C for periods up to 30 min. The response of the rock samples to the various thermal shocks was measured by the *modulus of elasticity* (*see* Chapter 3). This property is a useful indirect measure of a rock's resistance to weathering. It is directly related to *compressive strength* (again *see* Chapter 3), which in turn is correlated with the degree of weathering. High elasticity values suggest less weathered materials which are able to resist compressive stress, while low values indicate more highly weathered materials with low compressive strength.

The results of the experiments showed the effectiveness of very high temperatures as a weathering process. The first experiment showed the variety of rock responses, with granites being particularly vulnerable (*see* Table 6.5, column (a)). The second experiment showed that significant reductions in the modulus of elasticity occurred after as little as 1 min at 500°C (*see* Table 6.5, column (b)). Again, granites were particularly vulnerable.

Simulation experiments have also been carried out to examine what happens when wet rocks are exposed to the thermal shock of a fire. Rock outcrops are often damp before a fire event, especially when a fire is started by a lightning strike associated with thunderstorm activity. The results tend to show that the resistance of a rock to thermal shock under these conditions is determined by the physical properties that control water absorption.

6.4.4 Summary of the modifications needed to the simple model

We can now summarise the various complicating factors and you can see how they modify the simple model of insolation weathering outlined at the start of this section.

External factors

The main source of heat is solar radiation, whose variation during a 24 h period is modified by a number of factors. These may be **fixed**, where they include aspect, altitude and latitude (which influences air temperature), or **variable**, where they involve weather-related factors such as cloud, rainfall and windspeed. An important general consequence is that temperature fluctuations may be short term, and involve rapid change.

The contribution of bush fires is to provide very high temperatures for short periods of time.

Rock response

The *albedo* of a rock will influence the amount of effective solar heat received. The passage of this heat through a rock is affected by its *thermal conductivity*. This is generally low, which implies relatively high surface temperatures, high temperature gradients and expansion of the surface layers. On cooling there is contraction. The change in volume (or length) is determined by the *coefficient of thermal expansion*. Figure 6.5 shows the likely behaviour of an isotropic rock. When insolation occurs, surface layers will tend to expand, thus setting up tensile stress. When insolation ceases, surface layers will

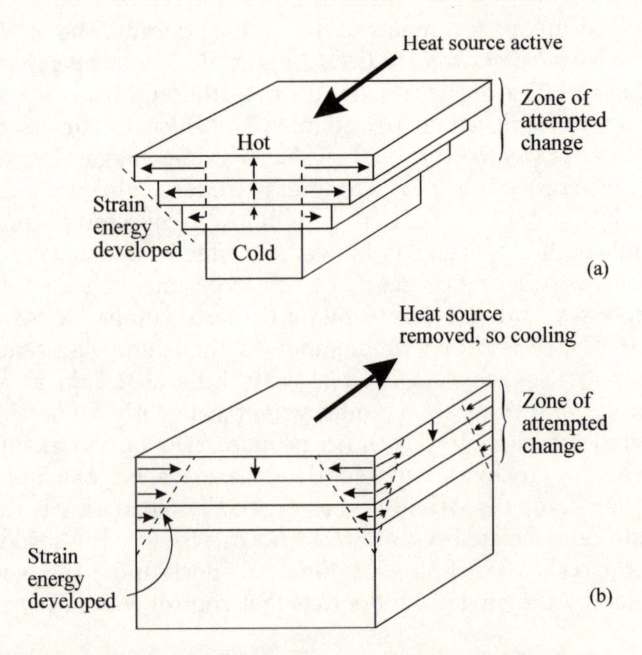

FIGURE 6.5 A simple block model, showing the development of tensional and compressive stresses. (a) Development of tensile stresses during the heating phase and (b) development of compressive stresses during cooling. (*Source*: Hall, K. and Hall, A. 1991: Thermal gradients and rock weathering at low temperatures: some simulation data. *Permafrost and Periglacial Processes* **2**, 103–12.)

tend to contract (especially when the air temperature is low) and so compressive stress will be developed.

You have, however, seen that at the individual mineral scale the coefficient of linear thermal expansion varies, both between minerals and with the direction of expansion for a given mineral. As many rocks consist of different minerals you can understand the complicated nature of the response to heating. For breakage finally to occur the elastic limit must be exceeded, and this also varies.

The physical principles behind insolation weathering are quite well established, but their application to specific circumstances raises many problems.

6.5 Pressure-release weathering

Joint sets parallel to the local land surface ('sheet' or 'dilation' joints) have been widely reported. They have been described for granitic terrains such as those of Dartmoor, England (*see* Chapter 11), New England, USA, and the Yosemite National Park region of the Sierra Nevada, USA. They have also been described for metamorphic areas, such as Jotunheimen, Norway.

The presence of sheet joints clearly weakens the bulk strength of rock. Most investigators believe that sheet joints are not original features, but form as a response to surface erosion. Their development may be seen as a type of mechanical weathering. There are two broad conditions under which such sheet jointing is produced.

6.5.1 Sheet jointing involving a change in compressive stress

A rock unit beneath the land surface experiences high compressive stress (*see* Chapter 3) because of the thickness and thus weight of rock above (*see* Fig. 6.6).

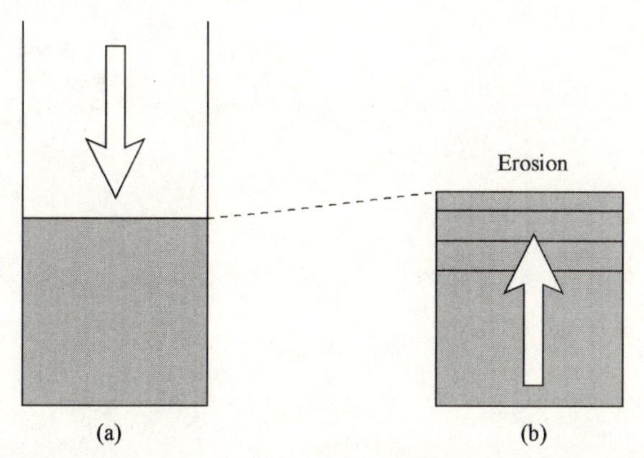

Figure 6.6 The effect of erosion on the balance between compressive and tensile stress. (a) High compressive stress on rock unit. (b) Compressive stress reduced by erosion; tensile stress dominates, and sheet jointing develops.

Box 6.7 The Griffith theory of crack development

In a classic study in 1921, Griffith proposed that the fracture of brittle material begins when the tensile stress σ_t at the tip of an (elliptical) crack provides sufficient energy to overcome the energy of the crack surfaces. This latter is associated with the breaking of interatomic bonds when the crack is formed. The relationship is:

$$\sigma_t \geq \left(\frac{2E\alpha}{\pi c}\right)^{1/2}$$

where α is the surface energy per unit area, E is Young's modulus (a measure of elasticity; *see* Chapter 3) and c is the half crack length.

Experimental work has shown that, for rock, a **fracture zone** with a large number of small cracks develops ahead of the main crack.

As surface erosion proceeds and the overlying rock is removed, the compressive stress on the rock unit is reduced and it will tend to expand in the direction of stress reduction. A threshold may be reached when the tensile stress (*see* Chapter 3) associated with expansion exceeds the tensile strength of the rock material. Failure is likely to begin with the extension of an initial crack (*see* Box 6.7) and continues with the development of sheet jointing parallel to the unloading surface. The process is most effective in brittle, crystalline rocks of high elasticity, such as granites and some metamorphic materials. It has also been described for massive (i.e. with widely spaced joints) sandstones.

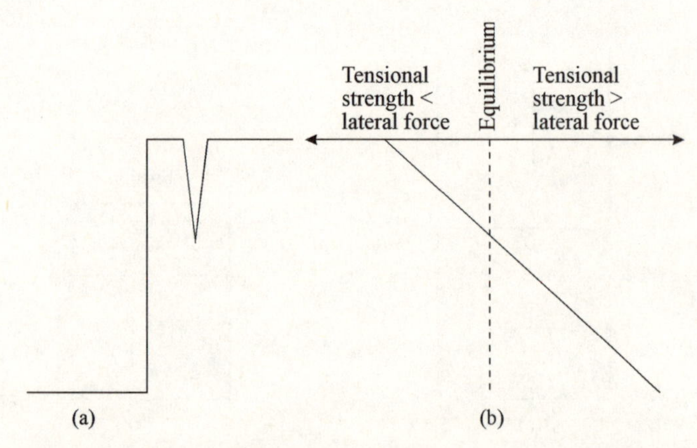

FIGURE 6.7 The relationship between crack development, tensional strength and tensional force. (a) Tension crack and (b) graph showing relationship between tensional force and tensional strength.

Box 6.8 Proposed mechanism of stress-corrosion cracking

Remember that a tectosilicate is basically made up of SiO_4^{4-} tetrahedral units linked at their corners by Si–O–Si bonds. The Si–O bond may be under mechanical stress, and if it can be broken, a microcrack in the silicate may extend. The figure below shows three steps in the reaction between water and an Si–O–Si bond under stress at a crack tip.

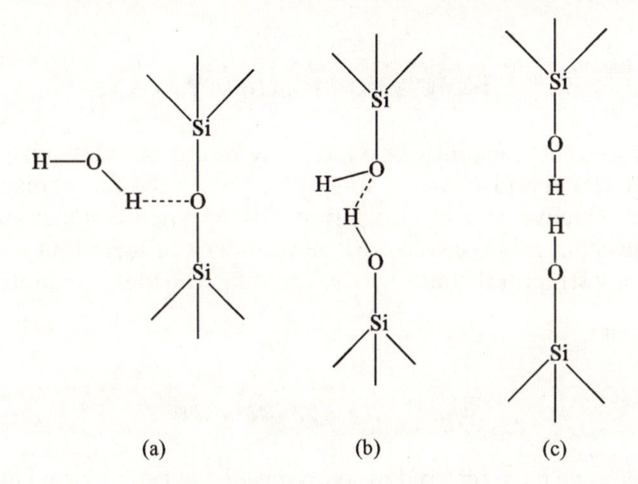

Now imagine a water molecule in the vicinity of this stress point (*see* diagram (a)). It will be attracted to the Si–O bond, so that its oxygen (O_w) bonds to the silicon via the oxygen lone pairs. One of its hydrogens can hydrogen bond to the bridging oxygen (O_b). The next step involves simultaneous proton and electron transfer, so that new bonds are formed, one between O_w and silicon, and one between hydrogen and O_b (*see* diagram (b)). The original bridging bond between O_b and silicon is destroyed. The final step is the breakage of the hydrogen bond between O_w and the transferred hydrogen to give an Si–O–H group on each fracture surface (*see* diagram (c)). The combination of a weakened bond and constant stress leads to microcrack extension.

(Based on Michalske, T.A. and Freiman, S.W. 1983: A molecular mechanism for stress corrosion in vitreous silicas. *Journal of the American Ceramic Society* **66**, 284–88.)

6.5.2 Sheet jointing involving mainly tensile stress

Observation in quarries in hard rock typically reveals the development of open joints parallel to and at the top of the side and back walls. A similar pattern has been described around cirques (armchair-shaped hollows formed by glacial action in mountain regions) in Jotunheimen, Norway. This variety of sheet jointing develops where the lateral (tensional) forces on a unit of rock exceed its resistance to tensile stress.

The balance of forces favours rupture at the top of a wall where resistance to tensile stress is lowest. Farther down, and due to the weight of overlying rock, tensile strength is greater than the lateral force and crack development ceases (Fig. 6.7). This could be several metres below the surface. The details of crack propagation are outlined in Box 6.7.

6.6 Stress-corrosion cracking

This is a process of crack growth which may be attributed to a combination of chemical action and physical stress at a point. The mechanism may be particularly effective at a crack tip in a silicate when water is present. In this case the chemical process is hydrolysis, which brings about the replacement of strong structural bonds by weaker links. The idea is explored further in Box 6.8.

Textual references

In this chapter we have referred to some specific investigations. Here are the references.

Goudie, A.S. 1974: Further experimental investigations of rock weathering by salt and other mechanical processes. *Zeitschrift für Geomorphologie* Suppl. **21**, 1–12.

Goudie, A.S., Allison, R.J. and McLaren, S.J. 1992: The relations between modulus of elasticity and temperature in the context of the experimental simulation of rock weathering by fire. *Earth Surface Processes and Landforms* **17**, 605–15.

Griggs, D.T. 1936: The factor of fatigue in rock exfoliation. *Journal of Geology* **44**, 781–96.

Hall, K. and Hall, A. 1991: Thermal gradients and rock weathering at low temperatures: some simulation data. *Permafrost and Periglacial Processes* **2**, 103–12.

Readings for further study

We have covered a lot of material in this chapter. As part of your study of a particular mechanical weathering process you may find it helpful to read first the discussion of that process in one or more geomorphological text books. You can then turn to the more detailed material in the readings below.

Allison, R.J. and Goudie, A.S. 1994: The effects of fire on rock weathering: an experimental study. In Robinson, D.A. and Williams, R.B.G. (eds) *Rock weathering and landform evolution*. Chichester: John Wiley, 41–56. Focuses on the effects of fire on rates of rock breakdown.

Matsuoka, N., Moriwaki, K. and Hirakawa, K. 1996: Field experiments on physical weathering and wind erosion in an Antarctic cold desert. *Earth Surface Processes and Landforms* **21**, 687–99. Focuses on the importance of salt weathering in the development of landforms.

McGreevy, J.P. 1981: Some perspectives on frost shattering. *Progress in Physical Geography* **5**, 56–75. A classic article, summarising many of the basic ideas about frost weathering.

Sperling, C.H.B. and Cooke, R.U. 1985: Laboratory simulation of rock weathering by salt crystallization and hydration processes in hot, arid environments. *Earth Surface Processes and Landforms* **10**, 541–55. A report on experiments designed to distinguish between the forces of crystal growth and hydration in salt weathering.

7

CHEMICAL WEATHERING

7.1 Weathering of rocks

When rocks are formed, the conditions will often involve high temperatures and pressures; there will probably be an absence of oxygen, carbon dioxide and water. Later, rocks will be out of equilibrium with their surroundings as they are exposed by erosion to the chemical weathering effects, particularly of water and dissolved gases. The action of these agents produces different results.

Water may dissolve materials completely (a process called *congruent solution*), for example limestone, to give gross features such as underground caverns. Water may also cause changes in aluminosilicates by removing a part of the material in solution (*incongruent solution*) to give rise to new materials, for example clay minerals.

Carbon dioxide in water will alter the pH of the water and may dramatically influence the solubility of, for example, minerals containing iron and aluminium as well as calcite and silica.

Oxygen may also be very important for weathering reactions especially oxidation, for example the formation of ferric oxide (Fe_2O_3) from ferrous oxide (FeO) in soils.

Chemical weathering processes can be divided into three categories:

- solution of ions and molecules;
- the production of new materials, e.g. clay minerals, oxides and hydroxides;
- the release of residual unweathered materials, e.g. quartz and gold.

The mechanisms which are components of these categories are hydration, solution, oxidation (and reduction), hydrolysis and complex formation via chelation. An example of the different categories is shown for a model silicate in Fig. 7.1.

Interlayer cations

Na^+K^+, $Ca^{2+}Mg^{2+}$ —hydrolysis, hydration, solution→ Solutions of $Na^+, K^+, Ca^{2+}, Mg^{2+}$

Aluminosilicate sheets (e.g. as parts of feldspars)

Al^{3+} and Si^{4+} in tetrahedra of Os —solution→ silicic acid (H_4SiO_4)

—hydrolysis, hydration→ secondary minerals, e.g. clays

Brucite and alumina
sheets and incorporated ions, e.g. Fe^{2+} (e.g. chlorite)

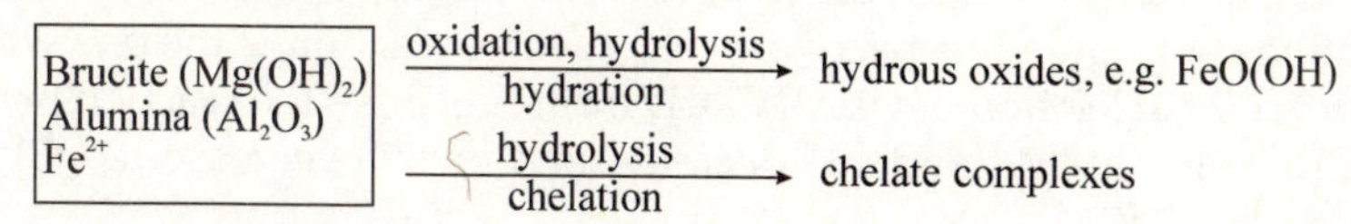

FIGURE 7.1 Diagrammatic representation of the chemical weathering processes which can break down a silicate.

In this chapter the principles of the main chemical weathering processes are explained and illustrated. Finally the chemical breakdown of some important minerals is described.

When considering the chemical changes which occur during weathering, you should realise that many factors are important. The first one to consider is the extent to which a weathering reaction proceeds to completion. Many weathering reactions will give complete conversion of parent material (reagents) to products while other reactions will give a measurable equilibrium. This behaviour is described by the equilibrium constant and the law of mass action which was briefly discussed in Chapter 2.

7.2 EQUILIBRIUM CONSTANT AND THE LAW OF MASS ACTION

When *reagents* come together they can combine to give *products*. For a *model reaction* we could write:

$$A + B \rightleftharpoons C + D$$

Note the arrows which show that $A + B$ can combine to give $C + D$. The reverse arrow shows that $C + D$ can combine to give $A + B$. A common example, which is of significance in weathering, is the hydrolysis of the carbonate ion (*see* section 7.10):

$$CO_3^{2-}(aq) + H_2O(l) \rightleftharpoons HCO_3^-(aq) + OH^-(aq)$$

At the start of a reaction no C and D will be present, so the only reaction is of A and B to give C and D. However, as soon as C + D is formed they can start to react to re-form A + B.

It is known that the rate of forward reaction, r_f, is

$$r_f = k_f[A][B]$$

where k_f is a constant for the forward reaction and [A] and [B] are the concentrations of A and B respectively. The forward reaction is the one which goes from left to right and in which A + B form C + D.

In the same way, the rate of reverse reaction, r_r, is

$$r_r = k_r[C][D]$$

where k_r is a constant for the reverse reaction and [C] and [D] are the concentrations of C and D respectively. The reverse reaction is the one which goes from right to left and in which C + D form A + B. As the reaction proceeds, r_f will decline as [A] and [B] decrease and r_r will increase as [C] and [D] increase.

At some point when equilibrium is reached,

$$r_f = r_r$$

so

$$k_f[A][B] = k_r[C][D]$$

This equation can be rearranged to:

$$\frac{k_f}{k_r} = \frac{[C][D]}{[A][B]}$$

k_f/k_r is called the equilibrium constant, K, for the reaction and, as shown above, occurs when the dynamic processes of forward and reverse reactions are equal. This means that A and B are being converted to C and D at the same rate as C and D are being converted to A and B, so the concentrations of A, B, C and D remain unaltered. It is important to emphasise that a chemical reaction does not go to some point at which everything stops. It goes to an equilibrium point at which the rates of the forward and reverse reactions become equal so that the concentrations of A, B, C and D are unaltered with time.

Now consider what happens if more A is added, e.g. more carbonate ion as in the example. The value of [A] increases and the rate of the forward reaction ($= k_f[A][B]$) increases and more C and D will be formed (note that K remains constant). In other words the reaction will move to the right. If we add more C, the value of [C] increases and more A and B would be formed.

The disturbing of the equilibrium and its re-establishment is an example of *Le Chatelier's principle* (1888):

'When a stress is applied to a system in dynamic equilibrium, the equilibrium adjusts to minimise the effect of the stress'.

So in this case the addition of A to the system results in the equilibrium changing to lower the concentration of A.

The relationship of concentrations of reagents and products in equilibrium can be written generally according to the law of mass action for the reaction:

$$aA + bB \rightleftharpoons cC + dD$$

as

$$K = \frac{[C]^c[D]^d}{[A]^a[B]^b}$$

An example could be for:

$$Al_2O_3 + 3H_2O \rightleftharpoons 2Al(OH)_3$$

for which:

$$K = \frac{[Al(OH)_3]^2}{[Al_2O_3][H_2O]^3}$$

(note that there is only one product in this example).

The expression of concentrations of A, B, C and D will depend on their nature. If the material is a solid, then by convention its concentration is expressed as 1. If the material is a gas it is expressed as a partial pressure; for example, oxygen in air has a partial pressure of 0.21 atmospheres, because oxygen has a proportion of 21 per cent by volume in the atmosphere. Carbon dioxide is about 330 parts per million by volume in the atmosphere, so its partial pressure is 0.00033 atmospheres. If the material is in solution the concentration is usually expressed as:

molarity = number of moles of solute in 1 litre of solution

or

molality = number of moles of solute per 1 kilogram of solvent

Box 7.1 Concentrations

Molarity is most commonly used to measure concentrations, but suffers from the disadvantage that the volume of a solution alters with change in temperature. For *molality* on the other hand, since the concentration is expressed per unit mass (1 kg) of solvent, there is no change with change in temperature. Most figures are usually quoted for temperatures of 25°C.

Between 0 and 100°C the volume of 1 g of water changes from 1.0013 to 1.0434 cm^3, so there is a change of approximately 4.3 per cent. Molality is therefore preferred on grounds of accuracy. However, the ease of making up solutions to a known volume means that molarity is most commonly used.

7.2.1 Activities

A further complicating factor in considering concentrations is that ions in aqueous solution are affected by the presence of each other. Cations will tend to be surrounded by anions, and anions will be surrounded by cations. Ions are not therefore interacting solely with water and will not behave as expected unless they are in very dilute solutions. They are said to behave *ideally* only when they are in extremely dilute solutions. In more concentrated solutions (i.e. not very dilute) the ions will exhibit behaviour expected of a lower concentration rather than the actual concentration. Instead of the term *concentration*, we use *activity*, which is a function of concentration calculated by multiplying the concentration by an activity coefficient:

$$a = \gamma c$$

where a is the activity of the ion, γ is the activity coefficient and c is the actual concentration.

In concentrated solutions γ is less than 1.

7.3 Spontaneous weathering reactions and thermodynamics

The oxidation of iron(II) oxide to iron(III) oxide by oxygen is a *spontaneous reaction*:

$$2FeO(s) + \tfrac{1}{2}O_2(g) \longrightarrow Fe_2O_3(s)$$

We can observe this from the red colour of iron(III) oxide (haematite) in soils, which is a well-known feature of warm environments and indicates that oxidation to iron(III) occurs spontaneously under appropriate conditions. The reverse reaction in which Fe_2O_3 decomposes to FeO and O_2 does not occur, i.e. it is not spontaneous. In order to explain why some reactions occur and others do not, you need to consider the principles of thermodynamics (energy changes).

You are probably familiar with the idea that reactions may be *exothermic* (when heat is given out) or in a few cases *endothermic* (when heat is taken in). These heat changes are called changes in *enthalpy* (H) but this is only one factor in the energy changes in a reaction. The other two factors are

- *Gibbs free energy* (G) which is the overall energy change;
- *entropy* (S) which is concerned with the energy changes associated with the degree of order (or 'randomness') of a system. For example, a solid (s) has a highly organised structure whereas a gas (g) has molecules moving in a very random fashion, so that a gas has high entropy and a solid has low entropy.

These energy factors are related by the equation:

$$\Delta G = \Delta H - T\Delta S$$

- the symbol Δ means 'change in', so ΔG is the change in Gibbs free energy for a chemical reaction;
- ΔH is the enthalpy change; T is the temperature expressed in degrees absolute or kelvin;
- ΔS is the change in entropy.

Spontaneous reactions occur when there is a decreasing value of G or in other words ΔG is negative.

Factors which will favour a negative value of the ΔG are

- a negative value of ΔH (an exothermic reaction) and
- a positive value of ΔS (increasing randomness in the reaction).

You should, however, notice from the equation $\Delta G = \Delta H - T\Delta S$ that if ΔH is positive (an endothermic reaction), it may still be offset by a positive value of ΔS (increasing randomness) to give a negative value of ΔG. Also, a negative value of ΔS (decreasing randomness in the reaction) can be offset by a negative value of ΔH (exothermic reaction) to give a negative value of ΔG.

In order to calculate the ΔG for a reaction, the values of ΔG for the formation of the reagents are added together and subtracted from the values of ΔG for the formation of the products:

$$\Delta G^{\circ}\text{ (reaction)} = \text{sum of } \Delta G_f^{\circ} \text{ of products} - \text{sum of } \Delta G_f^{\circ} \text{ of reagents}$$

where ΔG_f° for a compound is the energy change for its formation from the elements under standard conditions of 25°C and atmospheric pressure. For example, for the reaction

$$Fe(s) + \tfrac{1}{2}O_2(g) \longrightarrow FeO(s)$$
$$\Delta G_f^{\circ} = -251\ \text{kJ mol}^{-1}$$

Note that ΔG_f° for elements is zero. You should also note that many weathering reactions occur under *non-standard* conditions. We can now calculate ΔG° reaction for our example:

$$2FeO(s) + \tfrac{1}{2}O_2(g) \longrightarrow Fe_2O_3(s)$$

$$\begin{aligned}\Delta G^{\circ}\text{ (reaction)} &= \Delta G_f^{\circ}(Fe_2O_3) - (2\Delta G^{\circ}(FeO) + \tfrac{1}{2}\Delta G_f^{\circ}(O_2))^{*}\\ &= -742 - (-2 \times 251 + 0)\\ &= -742 + 502\\ &= -240\ \text{kJ}\end{aligned}$$

(*Remember that ΔG_f° for elements = 0.)

The negative value of ΔG° for this reaction means that it is energetically favoured and so is a *spontaneous reaction*.

Weathering reactions are often very complicated, so that the simple example considered above, where solid iron(II) oxide (FeO) reacts with gaseous oxygen to form solid iron(III) oxide (Fe_2O_3), would be unlikely to occur in a natural environment without the intervention of water in which oxygen is dissolved. In addition pure oxides would not normally be found. Nonetheless an appreciation of the basic principles will give you a greater understanding of weathering processes; for example, the oxidation of ferrous (Fe^{2+}) to ferric (Fe^{3+}) iron in this section was shown to be thermodynamically favoured, and this fits in with our observations of soil colour.

7.4 ΔG^o and equilibrium constant (K)

We can now use the ideas of ΔG to calculate values of equilibrium constant and so to find out the extent to which weathering processes will occur. The value of ΔG and K are related by the equation:

$$\Delta G^o = -RT \ln K \tag{7.1}$$

- R is a constant with the value $8.314\,\mathrm{J\,K^{-1}\,mol^{-1}}$;
- T is the temperature in degrees absolute or kelvin (for 25°C, T is $273.2 + 25 = 298.2\,K$; **note** that it is an unfortunate coincidence that the symbol for equilibrium constant is the same as that for kelvin - do not mix them up);
- $\ln K$ (ln is the natural logarithm) can be converted to $\log_{10} K$ by multiplying by 2.303:

$$\Delta G^o = -2.303RT \log_{10} K \tag{7.2}$$

From a value of ΔG^o the value of the equilibrium constant and hence the extent to which the reaction occurs can be estimated. For example, for the reaction

$$2FeO(s) + O_2(g) \longrightarrow Fe_2O_3(s)$$

we calculated earlier that ΔG^o (reaction) $= -240\,\mathrm{kJ}$.
Rearranging equation (7.2) gives

$$\log_{10} K = \frac{-\Delta G^o}{2.303RT}$$

Therefore

$$\log_{10} K = \frac{240\,000}{2.303 \times 8.314 \times 298.2}$$

that is

$$K = 10^{42.0339} = 1.081 \times 10^{42}$$

This large value of K indicates that the reaction will go virtually to completion.

While thermodynamics can indicate the likelihood of a reaction occurring, the results must be interpreted carefully. Consider, for example, the weathering of sodium feldspar to kaolinite by water saturated with carbon dioxide:

$$\underset{\text{sodium feldspar}}{2NaAlSi_3O_8(s)} + 9H_2O(l) + \underset{\text{carbonic acid}}{2H_2CO_3(aq)} \rightleftharpoons$$

$$\underset{\text{kaolinite}}{Al_2Si_2O_5(OH)_4(s)} + 2Na^+(aq) + 2HCO_3^-(aq) + \underset{\text{silicic acid}}{4Si(OH)_4(aq)}$$

Table 7.1 Free energies of formation

Formula	Name	ΔG_f^o (kJ mol^{-1})
$NaAlSi_3O_8(s)$	Sodium feldspar	−3698.7
$H_2O(l)$	Water	−237.2
$H_2CO_3(aq)$	Carbonic acid	−623.4
$Al_2Si_2O_5(OH)_4(s)$	Kaolinite	−3776.9
$Na^+(aq)$	Sodium ion	−261.9
$HCO_3^-(aq)$	Bicarbonate ion[a]	−587.0
$Si(OH)_4$	Silicic acid	−1308.8

[a] The modern name is hydrogencarbonate.

We can use the free energy of formation values given in Table 7.1 to calculate the ΔG^o value for the reaction:

$$\begin{aligned}\Delta G^o &= (-3776.9 - (2 \times 261.9) - (2 \times 587.0) - (4 \times 1308.8)) \\ &\quad -((-2 \times 3698.7) - (9 \times 237.2) - (2 \times 623.4)) \\ &= -10\,709.0 + 10\,779 \\ &= 69.1\,\text{kJ}\end{aligned}$$

The equilibrium constant for this reaction is

$$K = \frac{[Na^+]^2[HCO_3^-]^2[Si(OH)_4]^4}{[H_2CO_3]^2}$$

Remember that when solids are involved in a reaction their concentration is set at 1, so that the concentration of the sodium feldspar and of the kaolinite do not appear in the expression for the equilibrium constant. Also, because the concentration of water is very large and effectively constant it does not appear in the expression.

The value of the equilibrium constant can be calculated:

$$\Delta G^o = -2.303RT \log_{10} K$$

$$\log_{10} K = -\frac{\Delta G^o}{2.303 \times R \times T}$$

$$\log K = -\frac{69\,100}{2.303 \times 8.314 \times 298.2} = -12.1022$$

$$K = 7.9 \times 10^{-13}$$

In this case thermodynamics does not favour the reaction. Observation shows, however, that it occurs. If we consider water saturated with carbon dioxide in contact with the sodium feldspar, an equilibrium will be set up in which a small amount of the products will be formed in solution. If this solution moves away from the surface, fresh carbonated water comes into contact with the surface. More products will then be formed and so on until much, or all, of the sodium feldspar has been converted to kaolinite. The disturbing of the equilibrium and its re-establishment is an example of *Le Chatelier's principle* (*see* section 7.2).

7.5 Rate of reaction (kinetics)

In the previous section you saw that the extent to which a reaction will occur is determined by thermodynamic factors. We are now going to consider the rate at which a reaction moves towards equilibrium. The idea of 'rate' is explored further in Chapter 9. Even if a reaction is favoured thermodynamically it may proceed so slowly that effectively the reaction does not occur. In this case the term *quasi-equilibrium* is often used.

Generally the rates of weathering reactions are not as well understood as thermodynamic aspects. Even apparently simple processes, such as the rate at which calcite ($CaCO_3$) dissolves, have recently been shown to be highly complicated. While a full description of kinetics is beyond the scope of this book, a few basic principles can be described.

If we consider a model system,

$$A \longrightarrow B + C$$

the rate of reaction can be measured by the rate at which the concentration of A declines as the reaction proceeds, i.e.

$$-\frac{d[A]}{dt} = k[A] \qquad (7.3)$$

where [A] is the concentration at time t and k is the rate constant.

We could also have:

$$2A \longrightarrow B + C$$

and

$$-\frac{d[A]}{dt} = k[A]^2 \qquad (7.4)$$

where k is a different rate constant.

Equation (7.3) represents a *first-order rate equation* ($\alpha[A]$).

Equation (7.4) represents a *second-order rate equation* ($\alpha[A]^2$) since it depends on the square of the concentration of A.

If we plot the concentration of A against time for a *first-order reaction*, a curved plot will be seen as the concentration of A falls as the reaction proceeds (Fig. 7.2). If the reaction goes to completion then eventually [A] will fall to zero.

An important concept is the *half-life*, which is the time taken for half of the original amount of [A] to be converted. For first-order kinetics the half-life is constant. This is shown in Fig. 7.2 where $t_1 = t_2$. Kinetics are normally applied to chemical reactions, but the decay of radioactive elements also

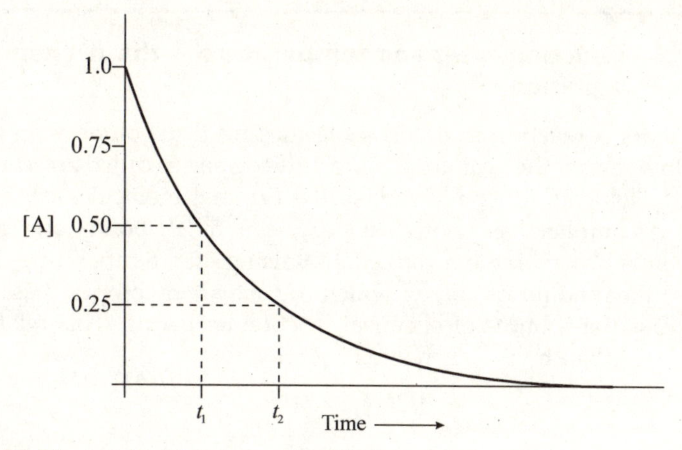

FIGURE 7.2 First-order kinetics and half-life.

obeys first-order kinetics and so can be used for purposes of dating in weathering studies (*see* Box 1.2).

7.6 EFFECT OF TEMPERATURE ON REACTION YIELDS AND RATES

The effect of temperature on the yields (the amount of products as a proportion of the amount of starting materials) of reactions can be predicted by the use of Le Chatelier's principle. If the temperature is raised, a reaction which is exothermic will have a lower yield, since effectively this lowers the amount of heat liberated. Similarly, a reaction which is endothermic will proceed further, as effectively this absorbs the extra heat. This means that exothermic reactions will tend to be dominant at the Earth's surface, but at the higher temperatures found within the Earth endothermic reactions may often be important.

The effect of an increase in temperature on reaction rates is to increase them. An approximate rule is that for every increase of 10°C the reaction rate is doubled. This may have significant consequences in weathering, e.g. weathering in the tropics, where the mean temperature is 20°C, is more rapid than in temperate climates where the average temperature is about 12°C. The quantitative relationship of temperature and rate is governed by the Arrhenius equation (*see* Box 7.2).

Ideas of activation energy (E_a) developed in Box 7.2 can be used in studies of weathering. Values of E_a can be obtained by measuring rates of reaction at different temperatures, and this technique has been used to measure the dissolution of silicate minerals. The rates of release of silica (SiO_2) were measured at different temperatures and the value of E_a was found for each mineral, given in Table 7.2.

Box 7.2 Reaction rates and temperature – the Arrhenius equation

In order for a reaction to occur, molecules need to collide with each other. In so doing they can form a short-lived species called an *activated complex*. This can either reassemble the original molecules or, if the activated complex has enough energy, can break bonds and form new bonds to give the product(s). Raising the temperature may increase the amount of energy which the activated complex has, and so increase the chances of chemical reaction occurring. This relationship is described by the Arrhenius equation:

$$k = A\,e^{-E_a/RT}$$

where k is the rate constant, A is the frequency factor, E_a is the activation energy, R is the gas constant and T is the temperature in kelvin.

The expression $e^{-E_a/RT}$ is the *Boltzmann expression* and represents the proportion of molecules which gain sufficient energy to go over the energy barrier to form products. This is shown in a *reaction profile*:

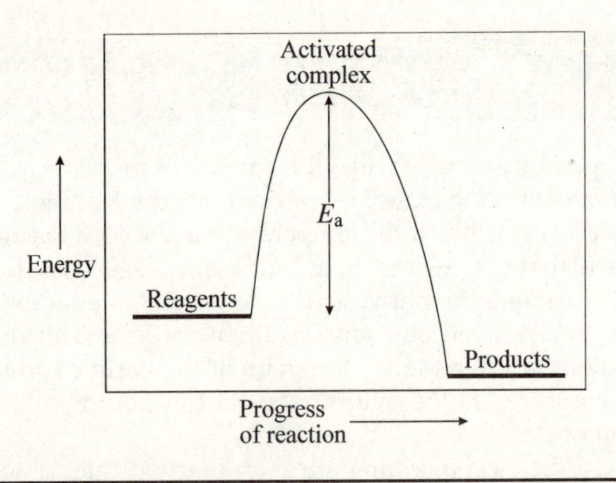

Table 7.2 Activation energies for dissolution of some silicate minerals

Mineral	E_a (kJ mol^{-1})
Diopside	38–81[a]
Enstatite	50
Potassium feldspar	38–82[a]
Albite	38–89[a]
Wollastonite	74
Quartz	77

[a] At various pH values.

The values are higher than for simple transport in solution, which is about 20 kJ mol^{-1}, but less than the values expected for removing ions from ionic crystals, which is about 80–400 kJ mol^{-1}. This suggests that adsorption on to the surface of the silicate aids the dissolution process, possibly by a catalytic effect.

7.7 HYDRATION

The hydration of simple ions was discussed in Chapter 2 and was shown to be important in helping to determine the solubility of ionic solids in water. Hydration of minerals can occur directly. In some cases there is formation of a hydrate with a definite number of molecules of water of crystallisation, for example

$$\underset{\text{anhydrite}}{CaSO_4(s)} + 2H_2O \rightleftharpoons \underset{\text{gypsum}}{CaSO_4 \cdot 2H_2O(s)}$$

The hydration of anhydrite to gypsum results in expansion and mechanical deformation. In areas where there is a horizontal hydration front, 'bubbles' of gypsum called *gypsum tumuli* may be created. These are domes of gypsum which have become separated from the underlying anhydrite by the forces of hydration. They are up to 2.5 m high and 10 m in diameter.

The relationship between anhydrite and gypsum by hydration or dehydration has consequences for weathering. Gypsum is the commonest of the sulphate minerals and is found in association with rocks such as clays and limestones as well as with evaporites generally.

Anhydrite is thought to be mainly a secondary mineral produced by the dehydration of gypsum in evaporite and other deposits. The well-defined formation of a hydrate is not the only form of direct hydration of minerals. The incorporation of water molecules into the layers of clay minerals leads to changes in volume due to swelling. Pressures greater than 300 MPa have been measured for the early stages of hydration of sodium montmorillonite. Such pressures can disintegrate quite hard rocks. The structure of silicates, of which clay minerals are a particular type, was discussed in Chapter 2 and the clay minerals themselves are described in Chapter 10.

7.8 SOLUTION

The factors which influence solubility of ionic solids were discussed in section 4.4. A good way of thinking about solubility is to imagine a beaker of water to which you add a finely divided solid (e.g. sodium chloride) while stirring the water. Initially the sodium chloride will dissolve and the solution is said to be *unsaturated* (i.e. more could dissolve) with respect to sodium chloride. If you keep on adding sodium chloride there will come a time when no more sodium chloride will dissolve and the solution is said to be *saturated* with respect to sodium chloride. This will occur at 25°C when 36 g of sodium chloride has been added to 100 g of water (called the solvent). If more sodium chloride is added no more will dissolve, however much sodium chloride is added. Sometimes *supersaturated* solutions may be found where the concentration of the compound in water exceeds its solubility. This usually occurs when saturated solutions are subjected to changing conditions, e.g. evaporation or cooling. Supersaturated solutions arise because there are no suitable surfaces or particles on which crystals can start to form. They will crystallise very rapidly once the process of crystallisation starts and the crystals present provide centres for it to continue. The process of crystal formation on suitable centres is called *nucleation*.

The term *ionic potential* is sometimes used in the literature on weathering, mainly as a measure of the ease with which an ion can be removed in solution from a compound (e.g. a mineral). Traditionally, ionic potential has been calculated by dividing the charge of the ion by its radius in angstrom units ($1\,\text{Å} = 10^{-8}\,\text{cm} = 10^{-10}\,\text{m}$). As examples:

$$\text{ionic potential of } K^+ = \frac{1}{1.33} = 0.8$$

$$\text{ionic potential of } Mg^{2+} = \frac{2}{0.65} = 3.1$$

Ions with low positive charge and a large ionic radius have a low value of ionic potential, e.g. K^+. This reflects a low charge density on the surface of the ion, so that it will only weakly attract negative ions in ionic solids and hence give low values of lattice energy. Water molecules will also be weakly attracted and the hydration energy will be low.

Ions with higher positive charge will tend to have a small ionic radius because the electrons are held more strongly. Both of these factors – high positive charge and small ionic radius – tend to give higher values of ionic potential. Positive ions with high ionic potential will attract negative ions more strongly in ionic solids (and give high lattice energy) and water molecules in aqueous solution (and give high hydration energy). You may notice that the effects of ionic potential work in the same direction on both lattice energy and hydration energy. It is not, therefore, easy to draw a simple conclusion about the effect of ionic potential alone.

What is observed in practice is that ions with low ionic potential are the main components of weathering solutions. The implication is that they are usually more mobile (i.e. more easily leached) in weathering processes. This is because they are more readily dissolved in water and are carried away. Values of ionic potential for some selected positive ions are shown in Table 7.3. You will note that, for example, the high value of Si^{4+} corresponds with low mobility of this ion in dissolution processes. The low value of Na^+ corresponds with the readily leached nature of this ion (*see* Chapter 10).

The solubility of a solid (solute) in water (solvent) can be expressed in terms of a *solubility product*; for example, for calcium sulphate ($CaSO_4$,

TABLE 7.3 Some values of ionic potential

Positive ion	Ionic potential (=charge/ionic radius)
Na^+	1.0
K^+	0.8
Ca^{2+}	2.1
Mn^{2+}	1.6
Fe^{2+}	2.6
Mg^{2+}	3.1
Fe^{3+}	5.6
Al^{3+}	6.6
Ti^{4+}	5.9
Mn^{4+}	7.4
Si^{4+}	9.8

anhydrite) we can write

$$CaSO_4(s) \rightleftharpoons Ca^{2+}(aq) + SO_4^{2-}(aq)$$

and the solubility product K_{SP} is given by

$$K_{SP} = [Ca^{2+}] \cdot [SO_4^{2-}]$$

For $CaSO_4$, $K_{SP} = 2.4 \times 10^{-5}$ and since $[Ca^{2+}] = [SO_4^{2-}]$,

$$[Ca^{2+}] \quad \text{or} \quad [SO_4^{2-}] = [CaSO_4]$$

The solubility of $CaSO_4$ is

$$\begin{aligned}\sqrt{2.4 \times 10^{-5}} &= 0.0049\,\text{mol dm}^{-3}\\ &= (40.1 + 32.1 + 64) \times 0.0049\\ &= 136.2 \times 0.0049\\ &= 0.67\,\text{g dm}^{-3}\end{aligned}$$

(Note: in this calculation moles are converted into grams by multiplying the number of moles of $CaSO_4$ by the molar mass of $CaSO_4$ (=40.1 + 32.1 + 64).)

For the mineral fluorspar (CaF_2) the solubility product is:

$$K_{SP} = [Ca^{2+}][F^-]^2 = 10^{-10.5}$$

Notice that in this case, because there are two F^- ions in the formula of CaF_2, the concentration of F^- is squared in the expression for the solubility product. Fluorspar has a very low solubility. In the case of both calcium sulphate and calcium fluoride the solubilities are very low, so the solutions will be very dilute. We can therefore use actual concentrations, but remember that for more concentrated solutions we need to use activities (*see* section 7.2).

Magnitudes of solubility in relation to the solubility product are affected by two factors, the common ion effect and ionic strength. The *common ion effect* is concerned with what happens when the concentration of one of the ions is increased. For example, if we add some potassium sulphate to a solution which is saturated in calcium sulphate, the concentration of sulphate ion $[SO_4^{2-}]$ will increase. In order to keep the solubility product constant, the equation

$$CaSO_4(s) \rightleftharpoons Ca^{2+}(aq) + SO_4^{2-}(aq)$$

will move from right to left, effectively decreasing the solubility of calcium sulphate. The same effect could be produced by adding the soluble salt, calcium nitrate ($Ca(NO_3)_2$), to the calcium sulphate solution. This increases $[Ca^{2+}]$ and moves the equation from right to left, but this time because of an increased concentration of Ca^{2+} rather than SO_4^{2-}. Weathering solutions are often complicated mixtures of ions and it may be difficult to estimate the solubility of compounds because of activity effects (*see* section 7.2) and common ion effects.

The *ionic strength* of a solution may also affect solubility. Ionic strength is concerned with the total number of ions present in a unit volume of solution and is defined by the equation:

$$I = \frac{1}{2} \sum m_i z_i^2$$

- I is the ionic strength,
- $\sum$ means 'the sum of',
- m_i is the concentration (in mol dm^{-3}) of the ion i
- z_i is the charge of the ion i.

For example, the ionic strength of a solution containing 0.01 mol dm^{-3} of $CaSO_4$ and 0.02 mol dm^{-3} of $MgCl_2$ is:

$$I = \frac{1}{2}((0.01 \times 4) + (0.01 \times 4) + (0.02 \times 4) + (0.04 \times 1)) = 0.1$$

As the ionic strength of a solution increases, the tendency for 'interference' between ions will increase. We can think of the interference as being caused by the hydrated ions of the added salt coming between the hydrated ions of the original salt and thus stopping it from precipitating. This lowers the activity of the ions so that the solubility of an ionic compound will tend to increase. Other things being equal, the ability of such a solution to dissolve a rock will be enhanced.

In very concentrated solutions the trend may be reversed with activity coefficients (*see* section 7.2) rising above 1, so that solubility will start to decrease. This, however, is unusual. The ionic strength of sea water is 0.7 and that of rivers and lakes is about 0.001. Alkaline lakes may have an ionic strength as high as 5. It has been estimated that the solubility of calcium sulphate in Mississippi river water, where $I = 2.2$, is increased by 21 per cent compared with its solubility in pure water. This favours transport in solution rather than precipitation.

The *temperature* effect on solubility occurs in a manner consistent with Le Chatelier's principle (*see* section 7.2). When a substance dissolves in water exothermically (i.e. heat is given out during solution) then higher temperatures will tend to give a lower solubility, since the giving out of less heat is helping to maintain the existing conditions. If the substance dissolves endothermically (heat is taken in) then solubility will increase with an increase in temperature.

7.9 ACIDS AND BASES

7.9.1 Acids

Acids and bases were introduced in Chapter 5 where you saw that an acid is a proton donor. The example chosen was hydrochloric acid. This is completely dissociated in aqueous solution and so is called a *strong acid.* Other examples of strong acids are sulphuric acid (H_2SO_4) and nitric acid (HNO_3). Some acids are said to be *weak acids* because they do not completely dissociate in aqueous solution. Many organic acids are weak. Ethanoic acid (CH_3COOH) is often chosen to illustrate the principles:

$$CH_3COOH(l) + H_2O(l) \rightleftharpoons CH_3COO^-(aq) + H_3O^+(aq)$$

Notice the way the equilibrium is drawn with a large arrow (↽) going from right to left and a small arrow going from left to right. This means that the equilibrium lies on the left-hand side of the equation, i.e. that there is a greater concentration of CH_3COOH than of CH_3COO^-. Using the law of mass action we can write:

$$K_a = \frac{[H_3O^+][CH_3COO^-]}{[CH_3COOH]}$$

(K_a is the acid dissociation constant).

Notice that the concentration of water is not included because it is very large and effectively constant.

At 25°C and atmospheric pressure

$$K_a = 1.76 \times 10^{-5}.$$

This means that only about 1 per cent of the ethanoic acid is dissociated in aqueous solutions. Strong acids are, however, almost completely dissociated and thus give higher $[H_3O^+]$ and so lower pH ($= -\log_{10}[H_3O^+]$) than organic acids. Ethanoic acid is found in the environment as a result of the action of bacteria. Other weak acids, for example humic acids in soil, carbonic acid (H_2CO_3) and hydrogen sulphide (H_2S), are also found and play a role in weathering processes (*see* Box 8.1).

7.9.2 Bases

These again may be strong or weak depending on the degree of dissociation. Sodium hydroxide (NaOH) is completely dissociated in water and so is a strong base:

$$NaOH(s) \xrightarrow{H_2O} Na^+(aq) + OH^-(aq)$$

Ammonium hydroxide (NH_4OH), which forms when ammonia is dissolved in water, is weakly dissociated and a weak base:

$$NH_3(g) + H_2O(l) \text{ (or } NH_4OH) \rightleftharpoons NH_4^+(aq) + OH^-(aq)$$

Here

$$K_b = \frac{[NH_4^+][OH^-]}{[NH_3]} = 2 \times 10^{-5}$$

K_b is the *base dissociation constant.*

Notice that the value of the constant is of the same order as for ethanoic acid, i.e. 10^{-5}, so ammonia is a weak base.

A number of important naturally occurring bases have very low solubility and therefore give low concentrations of OH^- in solution. Their status as weak bases is not due to an inability to dissociate.

Brucite ($Mg(OH)_2$) is an important example of the low solubility/weak base discussed above. In pure water the solubility is about $0.007\,g\,l^{-1}$ and gives rise to a pH of 10.4. The low solubility of brucite accounts for its widespread occurrence in nature. You should also note that it has two hydroxide (OH^-) groups and therefore dissociates in two stages:

$$Mg(OH)_2(s) \rightleftharpoons Mg(OH)^+(aq) + OH^-(aq) \tag{7.5}$$

$$Mg(OH)^+(aq) \rightleftharpoons Mg^{2+}(aq) + OH^-(aq) \tag{7.6}$$

For these dissociation steps the equilibrium constants are

$$K_1 = \frac{[Mg(OH)^+][OH^-]}{[Mg(OH)_2]}$$

Since $Mg(OH)_2$ is a solid, $[Mg(OH)_2] = 1$. Therefore

$$K_1 = [Mg^{2+}][OH^-] = 10^{-8.6} \qquad K_2 = \frac{[Mg^{2+}][OH^-]}{[Mg(OH)^+]} = 10^{-2.6}$$

Since K_2 is so much larger than K_1, Mg^{2+} rather than $MgOH^+$ usually dominates the occurrence of the magnesium species in aqueous solution. In acidic solutions the solubility of brucite will increase due to the removal of OH^- by the reaction

$$H_3O^+(aq) + OH^-(aq) \rightleftharpoons 2H_2O(l)$$

If the water is sufficiently acidic, then compounds other than $Mg(OH)_2$ may be more insoluble and so will occur as the solid phase in preference to $Mg(OH)_2$.

7.9.3 Silica (SiO_2) and silicic acid (H_4SiO_4 or $Si(OH)_4$)

Having studied the principles of acids and bases, we can now illustrate these ideas by applying them to the behaviour of an important common earth material which is weakly acidic. Silica is a solid compound which can form silicic acid by reaction with water:

$$SiO_2(s) + 2H_2O(l) \rightleftharpoons Si(OH)_4(aq)$$

Note that silica in the form of quartz is very slow to react with water, but freely formed (amorphous) silica is much more reactive.

Both $Si(OH)_4$ and H_4SiO_4 are used correctly as the formula for silicic acid in various text books. However, the formula $Si(OH)_4$ does indicate that four OH groups are attached to a central silicon, which is not so clearly demonstrated by writing the formula as H_4SiO_4.

Silica may eventually be formed as a result of the complicated weathering of aluminosilicates, e.g. feldspars and micas (*see* Chapter 10). The solubility of silica in water is affected by pH and it is not unusual for the solubility of SiO_2 to be exceeded, so that a precipitate of *amorphous silica* occurs, for example if the pH is lowered by the addition of water made acidic by dissolved CO_2. This precipitated silica is also sometimes described as a *hydrous silica* and is represented by the formula $SiO_2 \cdot nH_2O$. This amorphous silica (see section on silcretes) can then slowly expel water and form a more clearly crystalline material, a process called *ageing*. A sequence of changes can occur in which the first-formed material is opal, which is a hydrous cryptocrystalline form of cristobalite (cristobalite is a form of silica with a more open structure than quartz). Opal has tiny pores which contain water and so give rise to its characteristic milky appearance. It may have a water content of about 10 per cent by mass. A later stage is the formation of a group of varieties of silica called chalcedony, in which the silica is present as tiny crystals with submicroscopic pores. The varieties are chert, flint and jasper; agate is a material similar to chalcedony with bands of colour. The water content of chalcedony is lower, e.g. flint has about 1 per cent by mass of water. The increasingly condensed nature of the structures is shown by the increase in density from opal to quartz (*see* Table 7.4).

Quartz is very resistant to attack either chemically or physically, and thus is resistant to weathering processes except under the more severe conditions of humid tropical environments. The weathering of materials around quartz means that it is often concentrated by processes which form sediments and therefore occurs as a sand and subsequently as an important component of sandstones.

Amorphous silica, unlike quartz, has an appreciable solubility in water in the form of silicic acid. The solubility of silica can be pH dependent due to

Table 7.4 Density of different forms of silica

Variety	Density ($g\,cm^{-3}$)
Opal	2.1
Flint	2.6
Quartz	2.7

the possible dissociation of silicic acid:

$$H_4SiO_4(aq) \rightleftharpoons H_3SiO_4^-(aq) + H^+(aq) \tag{7.7}$$

$$H_3SiO_4^-(aq) \rightleftharpoons H_2SiO_4^{2-}(aq) + H^+(aq) \tag{7.8}$$

$$H_2SiO_4^{2-}(aq) \rightleftharpoons HSiO_4^{3-}(aq) + H^+(aq) \tag{7.9}$$

$$HSiO_4^{3-}(aq) \rightleftharpoons SiO_4^{4-} + H^+(aq) \tag{7.10}$$

The equilibrium constants for the above dissociations are:

$$(7.7) \quad K_1 = 10^{-9.7}$$

$$(7.8) \quad K_2 = 10^{-13.3}$$

$$(7.9) \quad K_3 = 10^{-9.9}$$

$$(7.10) \quad K_4 = 10^{-13.1}$$

We can contrast these very small values with that for

$$SiO_2\ (\text{amorphous}) + 2H_2O \rightleftharpoons H_4SiO_4$$

where $K = 10^{-2.7}$, which is several orders of magnitude greater than even K_1. This means that other than at pH above 9 the solubility of amorphous silica is not affected by pH. At pH above 9 the removal of H^+, for example from the right-hand side of equation (7.7), will promote further dissociation of H_4SiO_4 and hence increase the solubility of SiO_2. This is summarised in Fig. 7.3 where the concentration of silicic acid, expressed as an mg l^{-1} of silica, is plotted against pH. The species $H_3SiO_4^-$ becomes the most important one at pH = 9.7.

The concentration of silica in water is variable (*see* Table 7.5), with higher values generally found in rivers than in the sea. The values for sea water are often less because silica is removed from solution, e.g. by shellfish making

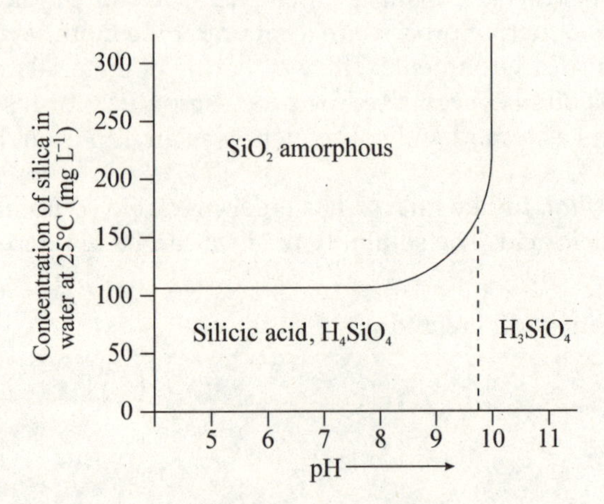

FIGURE 7.3 The solubility of amorphous silica (SiO_2).

TABLE 7.5 Concentration of silica (SiO_2) in natural waters

Source	Concentration ($mg\ l^{-1}$)
Mississippi river	12
Amazon river	11
Sea water	0.01–7

Source: Rankoma, K. and Saharna, T.H.G. 1950: *Geochemistry*. Chicago: University of Chicago.

their shells. This occurs at a faster rate than replacement by dissolution weathering.

7.10 HYDROLYSIS

It is important to distinguish between the words *hydration* (*see* section 7.7) and *hydrolysis*. Hydration does not involve reaction with water, but does involve an association between ions and water molecules. No bonds are broken in the process of hydration, whereas in hydrolysis one O–H bond of H_2O is broken (*see* Box 7.3).

An example of hydration would be

$$Mg^{2+} + 6H_2O \rightleftharpoons Mg(H_2O)_6^{2+}$$

in which the water molecules are attached to the magnesium ion (Mg^{2+}).

Hydrolysis may occur under neutral, acidic or basic conditions. An example is the weathering of silicon (Si^{4+}) from aluminosilicates, which occurs with splitting of water into H^+ and OH^- and the formation of silicic acid. This can be represented ideally by thinking of the silicon as an Si^{4+} ion, but we know that the silicon is actually at the centre of a tetrahedron of oxygen atoms to which it is covalently bonded:

$$Si^{4+} + 4H_2O \rightleftharpoons Si(OH)_4 + 4H^+$$

Hydrolysis occurs with compounds in which either the cations give rise to weak bases or the anions give rise to weak acids, or both the cation and anion give rise to weak bases and acids. Here are some examples. Calcium sulphate ($CaSO_4$) consists of calcium ions (Ca^{2+}) which give rise to a strong

Box 7.3 Use of the term '-olysis'

'-olysis' is used in a variety of chemical definitions, all of which imply the chemical reaction or 'breaking down' of a chemical compound. Two important weathering processes are

- hydrolysis: breaking down by reaction with water;
- acidolysis: breaking down by reaction with acid.

TABLE 7.6 Hydrolysis of salts

Type of salt	Example	pH of aqueous solution
Strong acid–strong base	$CaSO_4$	7
Weak acid–strong base	$CaCO_3$	>7 (basic)
Strong acid–weak base	$(NH_4)_2SO_4$	<7 (acidic)
Weak acid–weak base	$(NH_4)_2CO_3$	Close to 7 depending on the relative strengths of the acid or base

base ($Ca(OH)_2$) and sulphate ions (SO_4^{2-}) which give rise to a strong acid (H_2SO_4). Sodium carbonate ($NaCO_3$) is made up of a cation (Na^+) which gives rise to a strong base (NaOH) and an anion (CO_3^{2-}) which gives rise to a weak acid (H_2CO_3). The dissociation of sodium carbonate (Na_2CO_3) is

$$Na_2CO_3(s) \rightleftharpoons 2Na^+(aq) + CO_3^{2-}(aq)$$

The carbonate ion will then react with water (i.e. it is hydrolysed by water):

$$CO_3^{2-}(aq) + H_2O(l) \rightleftharpoons HCO_3^-(aq) + OH^-(aq)$$

This generates the hydroxide ion (OH^-) which makes the solution basic.

Ammonium sulphate ($(NH_4)_2SO_4$) is a good example of a weak base–strong acid system. The anion (SO_4^{2-}) can form a strong acid (H_2SO_4) while the cation (NH_4^+) gives rise to a weak base (NH_4OH). When ammonium sulphate is dissolved in water the following equilibria occur:

$$(NH_4)_2SO_4(s) \rightleftharpoons 2NH_4^+(aq) + SO_4^{2-}(aq)$$

$$NH_4^+(aq) + H_2O(l) \rightleftharpoons NH_4OH(aq) + H^+(aq)$$

The sulphate ion (SO_4^{2-}) remains in that form since sulphuric acid (H_2SO_4) is strong and so is fully dissociated in aqueous solutions. The ammonium ion is hydrolysed and has generated protons (H^+), so the solution is acidic. The overall situation is summarised in Table 7.6.

7.11 CARBONATION

Carbonation is concerned with the weathering effects of carbon dioxide (CO_2) in aqueous solution and interaction particularly with calcium carbonate. It is perhaps the Earth's commonest weathering mechanism and, on limestone, may give rise to solutional landscapes called **karst**. Solutions of carbon dioxide in water are weakly acidic. Water in equilibrium with the atmosphere, which contains about 0.03 per cent of carbon dioxide, has a pH of 5.6 due to the presence of dissolved carbon dioxide. Groundwater may be more acidic as the concentration of CO_2 can be 20–30 times higher than that in the atmosphere due to the generation of carbon dioxide by biological activity. For example, green plants respire approximately 40 per cent of their carbon dioxide through the roots. Bacteria and fungi are also prolific producers of carbon dioxide (*see* Chapter 8). This carbon dioxide is trapped in the soil structure and in temperate climates is commonly 0.1–3.5 per cent of soil air, but may be up to 11 per cent in tropical soils where higher temperatures give rise to greater biological activity and so greater production of carbon dioxide.

The important equation for solutions of carbon dioxide in water is

$$CO_2(g) + H_2O(l) \rightleftharpoons \underset{\text{carbonic acid}}{H_2CO_3(aq)} \tag{7.11}$$

This weak acid can dissociate further:

$$H_2CO_3(aq) \rightleftharpoons H^+(aq) + \underset{\text{bicarbonate ion}}{HCO_3^-(aq)} \tag{7.12}$$

Limestone (calcium carbonate, $CaCO_3$) is itself soluble to a limited extent in water (solubility in pure water at 25°C and 1 atm pressure = 100 mg L^{-1}):

$$CaCO_3(s) \xrightleftharpoons{H_2O} Ca^{2+}(aq) + \underset{\text{carbonate ion}}{CO_3^{2-}(aq)} \tag{7.13}$$

Hydrolysis of the carbonate ion can then occur:

$$CO_3^{2-}(aq) + H_2O(l) \rightleftharpoons HCO_3^-(aq) + OH^-(aq) \tag{7.14}$$

so that calcium-rich water (water which is in contact with limestone rocks) is basic.

The dissolution of calcium carbonate is affected by pH. Strong acid causes an obvious reaction with effervescence ('fizzing up') of carbon dioxide from the solution:

$$CaCO_3(s) + 2H^+(aq) \longrightarrow Ca^{2+} + H_2O(l) + CO_2(g) \tag{7.15}$$

Weaker acid will cause production of bicarbonate (also called hydrogen-carbonate):

$$CaCO_3 + H^+ \longrightarrow Ca^{2+} + HCO_3^- \tag{7.16}$$

The weathering of pyrites (FeS_2) causes production of sulphuric acid, which can cause the breakdown of limestone as shown above.

The addition of a base will lower the solubility of calcium carbonate due to the reversal of reaction (7.14), which will increase the concentration of carbonate ion and so drive reaction (7.13) from right to left.

Often limestone weathers by dissolution in water which contains dissolved carbon dioxide (carbonic acid, H_2CO_3), and then the reaction becomes:

$$CaCO_3(s) + H_2CO_3(aq) \rightleftharpoons Ca^{2+}(aq) + 2HCO_3^-(aq) \tag{7.17}$$

Reaction (7.17) partly explains the formation of underground caves. The formation of stalactites, i.e. precipitation of calcium carbonate, occurs by

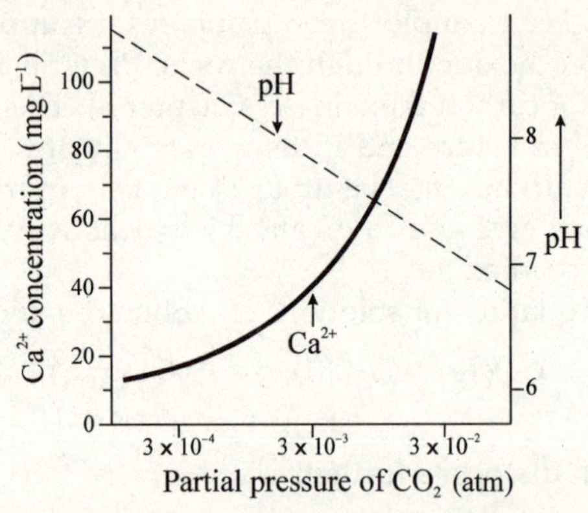

FIGURE 7.4 The solubility of $CaCO_3$ in pure water at 25°C with varying pressure of carbon dioxide.

the reverse of reaction (7.17) as follows.

1. Underground water equilibrates (i.e. comes into balance) with carbon dioxide at a pressure higher than that of the atmosphere.
2. This water dissolves calcium carbonate to a greater extent than would be the case with carbonated water at the surface of the Earth where carbon dioxide pressures are lower. The variation of solubility with carbon dioxide pressure is shown in Fig. 7.4.
3. This water, now supersaturated with calcium bicarbonate, on exposure to the atmosphere forms a drop at the top of a cave, and carbon dioxide comes out of solution due to the lower carbon dioxide pressure in the cave:

$$CO_2(aq) \rightleftharpoons CO_2(g)$$

Then some carbonic acid forms carbon dioxide in solution:

$$H_2CO_3(aq) \rightleftharpoons CO_2(aq) + H_2O(l)$$

4. The concentration of carbonic acid is now lower, so reaction (7.17) moves to the left and calcium carbonate precipitates, forming, in due course, a *stalactite* hanging from the roof of the cave.

Depending on the rate at which these changes occur, if equilibrium is not achieved, the drops may fall to the floor of the cave and produce a pillar growing upwards – a *stalagmite*.

A further example of carbonation occurs with the weathering of aluminosilicates by carbonic acid produced by the solution of atmospheric carbon dioxide in water, e.g. potassium feldspar ($KAlSi_3O_8$):

$$3KAlSi_3O_8(s) + 2CO_2(g) + 14H_2O(l) \longrightarrow KAl_3Si_3O_{10}(OH)_2(s) + 2HCO_3^-(aq) + 2K^+(aq) + 6H_4SiO_4(aq)$$

7.12 Complexes and Chelation

You should be familiar with the idea of chemical compounds, e.g. sodium chloride ($NaCl$), magnesium chloride ($MgCl_2$), calcium sulphate ($CaSO_4$) and magnesium hydroxide ($Mg(OH)_2$). In all of these examples the number of charges on positive ions is counterbalanced by the number of negative ions; for example, Na^+ (one positive charge) is counterbalanced by Cl^- (one negative charge). The charge on the Na^+ is called the *oxidation number* and is plus one (+1). In $MgCl_2$ the oxidation number of magnesium is +2. *The oxidation number is the same as the charge on ions in an ionic compound.*

We also have a *coordination number*, which is the number of nearest neighbours which the ion has. For magnesium chloride it is 2 as there are two Cl^- ions in the formula of the compound. Now consider $Mg(H_2O)_6^{2+}$ (*see* Fig. 7.5). This is found in aqueous solution and as an ionic species in the hydrate of magnesium chloride ($MgCl_2 \cdot 6H_2O$). In $Mg(H_2O)_6^{2+}$ the oxidation number of the magnesium is still +2 (remember that water molecules are neutral and so do not affect the charge). The coordination number (the number of nearest neighbours) of the magnesium is 6. We can now define the term complex:

a complex is formed when the coordination number of an ion is greater than the oxidation number.

You should note that the term 'coordination number' needs to be used with care. For example, consider sodium chloride – we saw in Chapter 1 that in the solid, NaCl, each Na^+ is surrounded by six Cl^-, so we say that the coordination number is 6, but for the purpose of determining whether NaCl is a compound or complex we consider just the formula of NaCl and say that one Cl^- is associated with Na^+ (the coordination number is 1). Thus the coordination number is the same as the oxidation number, so NaCl is a compound not a complex.

Natural examples are complicated so we have chosen a single compound, cobalt chloride ($CoCl_2$), to demonstrate some of the principles. Cobalt chloride is a compound which contains Co^{2+} and two Cl^-. The cobalt ion can also form a complex ion called hexaammine cobalt(III) ($Co(NH_3)_6^{3+}$). The species attached to the central atom (ammonia in our example) is called a *ligand*. This is a species which is attached to the central atom of a complex.

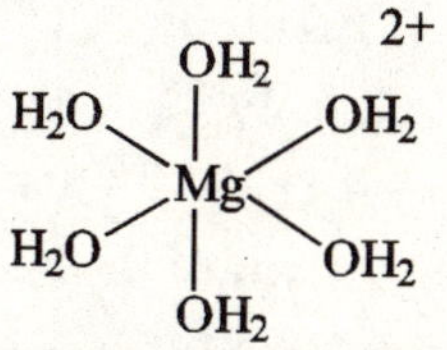

2+
OH_2
H_2O OH_2
Mg
H_2O OH_2
OH_2

Figure 7.5 The octahedral $Mg(H_2O)_6^{2+}$ ion found in aqueous solution and in $MgCl_2 \cdot 6H_2O$.

Box 7.4 The naming of complexes

The name of $Mg(H_2O)_6^{2+}$ is:

hexaaquomagnesium(II)

This looks very complicated, but is actually easy to understand if you break the name down:

'hexa'	means six;
'aquo'	is the word used for water in complexes;
'magnesium'	is the element;
'(II)'	is the oxidation number of the magnesium. The '(II)' could be omitted in this case since magnesium always occurs with a +2 oxidation number (unless it is the element for which the oxidation number is zero).

The name of $Co(NH_3)_6^{3+}$ is:

hexaamminecobalt(III).

The principles for naming this complex are as for $Mg(H_2O)_6^{2+}$. You should note that the name for ammonia in complexes is 'ammine' and that cobalt may commonly occur with a +2 or +3 oxidation number, so it is important to specify the oxidation number.

The ammonia molecule just has one point of attachment to the central atom (Co^{3+}) and so it is called a *monodentate ligand* (a 'one-toothed' ligand). Let us consider $NH_2CH_2CH_2NH_2$ (called ethylenediamine – 'ethylene' is the $-CH_2CH_2-$ group, 'di' means two and 'amine' refers to the $-NH_2$ groups). Ethylenediamine contains two nitrogen atoms and so it can attach itself to the central atom by two points of attachment; therefore it is called a bidentate ligand ('two-toothed' ligand). With Co^{3+}, ethylenediamine can form a complex as shown in Fig. 7.6.

The complex involving ethylenediamine is an example of a *chelate complex* (chelate means 'claw'). The ligand is called a chelate ligand because it grips the metal ion with more than one point of attachment. The significance of this is that chelate complexes are more stable than complexes which involve only monodentate ligands. Some ligands form exceptionally stable complexes;

FIGURE 7.6 Complexes of Co^{3+}.

for example, ethylenediaminetetraacetic acid or EDTA (which is widely used to assess the calcium content of natural waters) has six points of attachment to a central atom. There are two nitrogen and four oxygen atoms from the acetic acid groups:

$$\begin{array}{ccc} HO_2C{-}CH_2 & & CH_2CO_2H \\ & \diagdown\; N{-}CH_2{-}CH_2{-}N \;\diagup & \\ & \diagup \qquad\qquad\qquad \diagdown & \\ HO_2C{-}CH_2 & & CH_2CO_2H \end{array}$$

An important example of chelation occurs during the release and mobilisation of iron(III) (Fe^{3+}) and aluminium(III) (Al^{3+}). The complex formation occurs with the fulvic acids (*see* Box 8.1). These metals are then transferred down the soil profile. Fulvic acids form an acid-soluble fraction which is found in humus together with humic acids and humin (an insoluble residue of humus). They are mixtures of large molecules which contain a range of structural features. The most important are the carboxylic acid groups ($-CO_2H$) and hydroxyl groups ($-OH$). These molecules can form chelate complexes with Fe^{3+} and Al^{3+}. These chelate complexes are soluble and thus can move in solution down through the soil until they are precipitated as oxides or hydroxides, for example by the presence of Ca^{2+} or Mg^{2+} or by the presence of bacteria which break down the complexes. The chelation process may explain why certain metals are more mobile in podzolisation than might be expected from pH considerations alone.

7.13 Redox reactions

Redox weathering reactions commonly occur in aqueous solution. The principles of redox reactions were discussed in section 5.2. All redox reactions require something to be reduced and something else to be oxidised. For weathering processes it may be only one of the products which is of interest, so the treatment here is divided into 'oxidation' where oxidised products are important and 'reduction' where reduced products are important.

7.13.1 Oxidation

This usually occurs with oxygen as the oxidising agent, which is normally dissolved in water. The most commonly oxidised material is iron, which is converted from ferrous (Fe^{2+}) to ferric iron (Fe^{3+}). The yellowish brown to red colour of tropical soils is typically due to weathering in which iron is oxidised. There are two oxidation mechanisms:

- oxidation of Fe^{2+} in the mineral and subsequent changes;
- release of Fe^{2+} and its rapid oxidation to Fe^{3+}.

The oxidation of Fe^{2+} in the mineral can be illustrated by biotite, which is one of the trioctahedral micas. It has the formula

$$[(Mg^{2+}, Fe^{2+})_3(Si_3Al)O_{10}(OH)_2]^-K^+$$

which can be compared with *muscovite*

$$[(Al_2)(Si_3Al)O_{10}(OH)_2]^{-}K^{+}$$

which is a dioctahedral mica.

Biotite is much more easily weathered than muscovite, largely because oxidation of Fe^{2+} to Fe^{3+} alters the balance of charge in the structure, leading to the formation of other aluminosilicates and release of the ferric oxide. In contrast, a lack of iron in muscovite means that the oxidation mechanisms cannot work and so the mineral is more resistant to weathering. The second mechanism which involves the initial release of Fe^{2+} is illustrated by the hydrolysis of olivine, which causes loss of ferrous oxide. The ferrous oxide is then rapidly oxidised to ferric oxide in its hydrated form, goethite (FeO(OH)):

$$\underset{\text{olivine}}{3(MgFeSiO_4)} + 2H_2O \longrightarrow \underset{\text{serpentine}}{H_4Mg_3Si_2O_9} + SiO_2 + 3FeO$$

$$2FeO + \tfrac{1}{2}O_2 + H_2O \longrightarrow \underset{\text{goethite}}{2FeO(OH)}$$

Fayalite provides a further example of the same mechanism (hydrolysis followed by oxidation):

$$\underset{\text{fayalite}}{Fe_2SiO_4} + 4H_2O + 4CO_2 \longrightarrow 2Fe^{2+} + 4HCO_3^{-} + H_4SiO_4$$

$$4Fe^{2+} + O_2 + 4H^{+} \longrightarrow 4Fe^{3+} + 2H_2O$$

The Fe^{3+} is eventually found in mineral form, e.g. FeO(OH) (as above), $Fe(OH)_3$ or Fe_2O_3.

Another element which undergoes oxidative weathering is manganese (*see* Chapter 10), as it can readily exist in the +2 (Mn^{2+}) or +4 (Mn^{4+}) oxidation state:

$$\underset{\text{rhodocrosite }(Mn^{2+})}{MnCO_3} + \tfrac{1}{2}O_2 + H_2O \longrightarrow \underset{\text{pyrolusite }(Mn^{4+})}{MnO_2} + H_2CO_3$$

7.13.2 Reduction

Weathering processes involving reduction of the important species are less significant than oxidation reactions. It frequently occurs in an anaerobic (air/oxygen-free) environment such as that found in waterlogged soils. The characteristic colour of such soils is green grey and is due to the reduced form of oxides of iron in oxidation state +2. This contrasts with the red or yellow colour of compounds of iron in its +3 oxidation state.

Organic matter commonly functions as a reducing agent in weathering processes (*see* Chapter 8). The organic compounds are themselves oxidised to form carbon dioxide and water or to form new organic compounds. Sulphate ions (SO_4^{2-}) may be reduced by bacteria which use the oxygen in the sulphate to oxidise organic material. A typical reduction of sulphate is via the oxidation of an organic alcohol to an organic acid (carboxylic acid):

$$\underset{\text{organic alcohol}}{2RCH_2OH} + SO_4^{2-} \longrightarrow \underset{\text{organic acid}}{2RCOOH} + 2H_2O + S^{2-}$$

where R represents an organic group, e.g. CH_3-.

The sulphide ions can then go on to react in various ways:

- hydrolysis may occur to form H_2S:

$$2H^+ + S^{2-} \longrightarrow H_2S$$

- precipitation may occur with reduced forms of iron (Fe^{2+}) or manganese (Mn^{2+}):

$$S^{2-} + Fe^{2+} \longrightarrow FeS$$

$$S^{2-} + Mn^{2+} \longrightarrow MnS$$

Reduction processes which result ultimately in the formation of metals are discussed in Chapter 10.

7.14 Cation exchange

Fine soil particles produced by weathering processes are particularly important in providing a reservoir of exchangeable cations and anions, which may be involved in further weathering processes. The particles which are colloidal in nature (<0.2 μm diameter) may be either clay minerals or humus. Each particle can be thought of as a central unit which carries many negative charges on its surface, and individual positive exchangeable ions attached to the surface (*see* Fig. 7.7). Exchangeable cations are more common than exchangeable anions and may consist of, for example, H^+, K^+, Ca^{2+}, Mg^{2+} and Al^{3+}.

The negative charges on the colloidal particle arise in different ways for clay minerals and humus.

In clays the replacement of Al^{3+} for Si^{4+} at the centre of a tetrahedron of oxygens causes an excess negative charge of one. This negative charge may be satisfied by cations absorbed between the layers of a clay mineral as well as by cations adsorbed on the surface (exchangeable cations; *see* Fig. 7.8). The inter-layer cations will only normally exchange if the inter-layer spacing is greater than 1.4 nm.

In humus the negative charges are provided by the acidic organic groups, e.g. $-CO_2H$ and phenolic OH groups (*see* Chapter 8). These acidic groups can ionise to give negative sites as shown in Fig. 7.9.

These cations may be exchanged if water containing other cations is passed through the soil or weathering material:

$$\boxed{\text{colloid}}\,Ca^{2+} + 2H^+(aq) \longrightarrow \boxed{\text{colloid}}\begin{matrix}H^+\\H^+\end{matrix} + Ca^{2+}(aq)$$

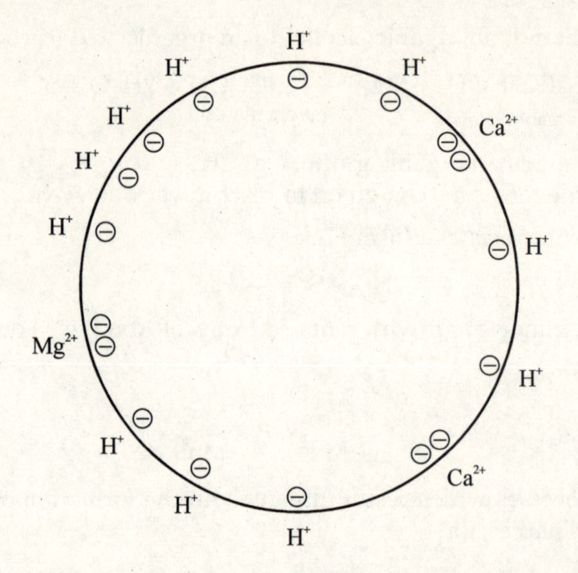

FIGURE 7.7 Simplified colloidal soil particle showing exchangeable cations attached.

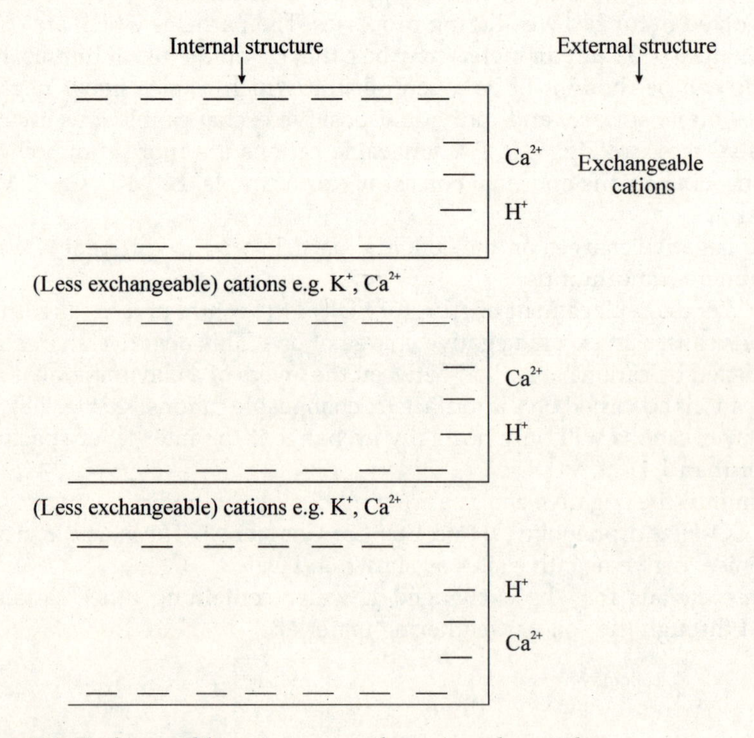

FIGURE 7.8 Exchangeable cations on a clay mineral particle.

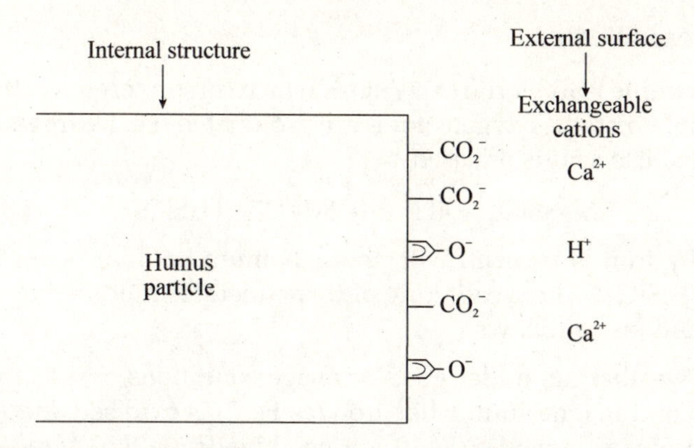

FIGURE 7.9 Exchangeable cations on a humus particle.

In this example, exchangeable Ca^{2+} is replaced by protons in acidic water, resulting in release of the Ca^{2+} ions into water and attachment of the two H^+ ions to the colloid surface.

Different soils have different amounts of exchangeable cations and this leads to the idea of *cation exchange capacity* (CEC), which is the concentration of cations in milliequivalents per 100 g (=centimoles of positive charge per kilogram). Remember that ions may carry single (e.g. H^+ or K^+) or double (e.g. Ca^{2+} or Mg^{2+}) or triple (e.g. Al^{3+}) or quadruple (e.g. Si^{4+}) charges.

CEC values in centimoles per kilogram of positive charge are typically:

Sandy soils	>10
Loam soils	10 (not usually much higher) to 30
Clay soils	up to 60

The presence of these cations may have an effect on the composition of soil solution, which may in turn modify weathering processes in the regolith. They may also modify precipitation processes (*see* Chapter 10).

7.15 CASE STUDIES OF SILICATES

Much of this chapter has been concerned with the processes by which rocks and minerals are weathered. In this section you will build on this knowledge of processes to examine the weathering of silicates. The products of this weathering are discussed in Chapter 10. The weathering of minerals by aqueous solutions is influenced by two main factors:

- the composition and lattice structure (including defects) of the mineral;
- the nature and behaviour of the weathering process.

We shall focus on the silicate minerals since these make up over 90 per cent of the Earth's crust and comprise about 75 per cent of the rocks exposed at the land surface. The main types of silicates are listed in Table 2.2.

7.15.1 Nesosilicates

A simple example is forsterite (Mg_2SiO_4), which can undergo weathering by hydrolysis in which the oxygen atoms are protonated (i.e. hydrogen ions are added) and silicic acid is released:

$$Mg_2SiO_4 + 4H^+ \longrightarrow 2Mg^{2+} + H_4SiO_4$$

More usually iron is present, with the maximum amount being found in olivine ($MgFeSiO_4$). The weathering of this mineral is modified by environmental conditions as follows.

- Tropical weathering, under good drainage conditions, results in the loss of silicon and magnesium, while iron (as Fe^{2+}) is oxidised and redistributed typically as hematite (Fe_2O_3) or goethite ($FeO(OH)$) along any fractures. The result is a porous neoformation with a structure that reflects the original pattern of fractures.
- Under less extreme conditions only magnesium (Mg^{2+}) is removed and an iron–silicon colloid fills up a framework of iron oxides. Eventually the silicon is removed and only the ferruginous network is left.
- Under moderate conditions most of the components are recombined to form an iron-bearing smectite, which may occur in bands along the original fractures.

7.15.2 Inosilicates (single and double chains)

The chemical weathering of minerals in this class, mainly pyroxenes and amphiboles, is guided by their good cleavage.

Under conditions of slow leaching, smectites are formed in cleavages and any fissures. Subsequently, and under stronger leaching conditions, any pyroxene and amphibole residues dissolve congruently, leaving iron-stained smectite. Tropical weathering and good drainage lead to the total loss of Ca^{2+}, Mg^{2+} and Si^{4+}, resulting in a neoformation consisting mainly of oxohydroxides, e.g. goethite.

7.15.3 Phyllosilicates

The commonest primary phyllosilicates are the micas, e.g. biotite and muscovite. Serpentine and chlorite are less common. Very good cleavage is characteristic, and provides important pathways for weathering. *Biotite* provides the best example of phyllosilicate weathering. Several alteration sequences may occur.

- Conversion to *chlorite*, which may be brought about in two ways. A brucite sheet may replace the potassium ion interlayer (K-interlayer) of biotite; alternatively, the loss of tetrahedra from one biotite layer may reduce it to a brucite-like layer so that two biotites become one chlorite.
- Conversion to *vermiculite*, which occurs mainly by the replacement of K^+ by hydrated cations.

- Further evolution towards the clay minerals kaolinite and gibbsite, or to the iron oxide, goethite, may occur. The conversion of mica to gibbsite by a hydrolysis mechanism is

$$2KAl_3Si_3O_{10}(OH)_2 + 18H_2O + 2H^+ \longrightarrow 3(Al_2O_3 \cdot 3H_2O) + 6H_4SiO_4 + 2K^+$$

7.15.4 Tectosilicates

Feldspars are the dominant tectosilicates, and indeed are the most important single group of rock-forming silicate minerals. They show very good cleavage but initial weathering typically takes place along fracture surfaces.

The chemical equations for the main alteration processes are quite straightforward:

- to *kaolinite*, involving hydrolysis:

$$\underset{\text{orthoclase}}{4KAlSi_3O_8} + 22H_2O \longrightarrow \underset{\text{kaolinite}}{Al_4Si_4O_{10}(OH)_8} + 8H_4SiO_4 + 4K^+ + 4OH^-$$

- to a *mica* involving hydrolysis:

$$3KAlSi_3O_8 + 12H_2O + 2H^+ \rightleftharpoons \underset{\text{mica}}{KAl_3Si_3O_{10}(OH)_2} + 6H_4SiO_4 + 2K^+$$

However, the mechanisms of alteration are rather more complicated.

Experimental studies have shown that weathering solutions adjacent to feldspar surfaces quickly show a high pH and the presence of basic cations. This evidence suggests that the first stage in weathering is the penetration of available protons, and their exchange with cations, particularly at external sites. A result is an excess of negative charge around the oxygen atoms just inside the structure. Forces of repulsion increase, and the way is opened to the interior cations. However, the Si–O–Al framework is still unaltered. For this to be broken, the penetrative H^+ ions must link to the O^{2-}. In addition OH^- ions from water bond with silicon and aluminium, to yield $Si(OH)_4$ and $Al(OH)_3$. In these ways feldspar is dismantled.

READINGS FOR FURTHER STUDY

There are few up-to-date text books devoted to chemical weathering itself. The topic is often covered in modern text books on geochemistry.

Atkins, P. and Jones, L. 1997: *Chemistry, molecules, matter and change*, 3rd edn. New York: W.H. Freeman. We have already referred you to this book and it provides a good background for the basic chemistry which underlie weathering processes.

Faure, G. 1991: *Principles and applications of inorganic geochemistry*. New York: Macmillan. This is a good example of a geochemical text in which various aspects are covered, often at an advanced level.

Ford, D. and Williams, P. 1989: *Karst geomorphology and hydrology*. London: Unwin Hyman. This book is for advanced undergraduates and graduate students. Not only is it an authoritative account of karst landscapes – sinking streams, caves, enclosed depressions, fluted rock outcrops and large springs – but it also covers ideas of, for example, dissolution, carbonation and chemical equilibria.

Ollier, C. 1984: *Weathering*, 2nd edn. London: Longman. This classical book contains a short description of chemical weathering processes.

Yatsu, E. 1988: *The nature of weathering. An introduction*. Tokyo: Sozosha. This excellent text book is comprehensive in its coverage of the literature of chemical weathering, but gives less emphasis to the underlying principles.

8

BIOLOGICAL WEATHERING

The immediate causes of rock and mineral breakdown are physical and chemical processes. The agents of these mechanisms are often, however, living organisms whose zone of activity is called the *biosphere*. Their contribution is sufficiently distinctive for biological (or biotic, meaning 'relating to life') weathering to be recognised as a separate category of material breakdown, with biophysical and biochemical processes as natural subdivisions. Some general ideas are discussed first and then the work of the various agents is reviewed.

8.1 Basic ideas

8.1.1 Variety and contribution of biological agents

Important biological agents include bacteria (*see* Box 8.2), lichens, algae, fungi (*see* Box 8.5), plant roots, and organic matter in a state of decay. The work of each of these agents is discussed separately for convenience, but it is most important to appreciate that in the real world they may drive similar processes. In addition, the details of the various mechanisms are still in many cases obscure.

Organisms assist rock and mineral breakdown in two main ways. They may exert physical stress on rock through, for example, the expansion and contraction of plant organic tissue during wetting and drying. As part of their life processes they may emit substances which include carbon dioxide, complexing agents and a range of organic and inorganic acids (*see* Box 8.1), as well as protons and electrons. These substances and particles trigger a range of processes, including oxidation, reduction, complexation (the formation of complex molecules), chelation and precipitation.

8.1.2 Distribution and density

Biological agents are found in a wide variety of environments, including the most severe. Bacteria have been recorded in hot springs at 90°C, in the

TABLE 8.1 Distribution of some biological agents (microorganisms) within the weathering layers of an olivine rock

Weathering layer		Colony-forming units (cfu g^{-1})		
Description	**Depth (cm)**	**Actinomycetes**	**Other bacteria**	**Fungi**
Surface	0–1	3.9×10^4	1.1×10^5	3.5×10^2
Subsurface	1–2.5	1.0×10^4	7.6×10^4	1.2×10^3
	2.5–4	2.9×10^5	2.9×10^3	8.7×10^2

Source: Webley, D.M., Henderson, M.E.K. and Taylor, J.F. 1963: *Journal of Soil Science* **14**, 102–12.

oceans to depths of 5000 m, in the wetted pores of crustal rocks to a depth of about 3 km and under rock surfaces in hot deserts. Lichens have been studied in the Antarctic. Algae have been investigated on rock outcrops (nunataks) above an icefield in Alaska. In less harsh environments the upper layers of the regolith (*see* Box 10.2), notably the root zone (or *rhizosphere*), are sites of intense biological activity.

The wide distribution of biological agents is enhanced by the high density of some varieties. On average (expressed as cfu, or colony-forming units, per gram dry weight), 10^6 actinomycetes (*see* Box 8.2), 10^7 other bacteria, 10^5 fungi and 5×10^4 algae are found in the top layers of cultivated soil. Horizontal and vertical distributions naturally vary, perhaps by magnitudes of 10 or more, depending on various factors. A similar concentration is found for rocks (*see* Table 8.1).

8.1.3 Importance of organic and inorganic acids

The production of acid is perhaps the most important contribution that biological agents can make to the breakdown of rocks and minerals. While both inorganic and organic types are involved, the latter is by far the more important in weathering reactions.

Inorganic acids largely arise from oxidation. For example, sulphuric acid (H_2SO_4) can be formed by the bacterial oxidation of sulphur compounds. The oxidation of ammonium ions (NH_4^+) by bacteria yields protons (H^+), as discussed further in section 8.2.1.

Organic acids (*see* Box 8.1) are emitted by many biological agents, e.g. lichens, bacteria, filamentous fungi, algae and cyanobacteria, as part of their life processes. They may also arise as by-products of organic decomposition. Their contribution to weathering partly reflects their solubility in water, and is also affected by the type of weathering mineral. Olivine, for example, responds much more rapidly to organic acid attack than to weathering under more sterile conditions (*see* section 9.2.1). While the reaction of minerals to organic acids is not well understood, some general principles of mineral response may be stated.

Typically, monovalent cations in the weathering mineral are replaced by protons (H^+) and washed out by water. Di- and trivalent cations are made

Box 8.1 Organic acids

These are relatively weak acids compared to the stronger mineral acids such as sulphuric acid. They are derived from various organic sources.

Simple organic acids

Some organic acids are simple compounds with well-defined molecules. They usually contain the carboxylic acid group ($-CO_2H$). This ionises to give a proton:

$$-CO_2H \longrightarrow -CO_2^- + H^+$$

Examples include the following acids.

Acid	Formula	Source
Formic (methanoic)	HCO_2H	Derived especially from ants
Acetic (ethanoic) Lactic Oxalic Gluconic	CH_3CO_2H $CH_3CH(OH)CO_2H$ $(CO_2H)_2$ $CH_2OH(CHOH)_4CO_2H$	Derived from the action of bacteria. **Oxalic** acid is an important exudate of lichens and fungi and **gluconic** acid of algae
Citric	$C(OH)(CO_2H)(CH_2CO_2H)_2$	Produced by fungi and lichens, as well as by bacteria
Malic	$C(OH)(CO_2H)(CH_2CO_2H)_2$	Present in the juice of many sour fruits and some plants

Complicated organic acids

These are mixtures of large complicated molecules found in soil and arising from decaying organic matter, especially plants. It is not possible to write down one formula for these acids. *Fulvic acid* is more soluble than *humic acid*, and this is how they are distinguished. They contain two types of acidic group, the carboxylic acid group as discussed above, and the phenolic group.

Phenol itself has the structure

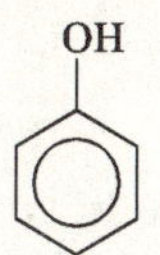

Box 8.1 *Continued*

and can ionise to produce protons and is therefore acidic:

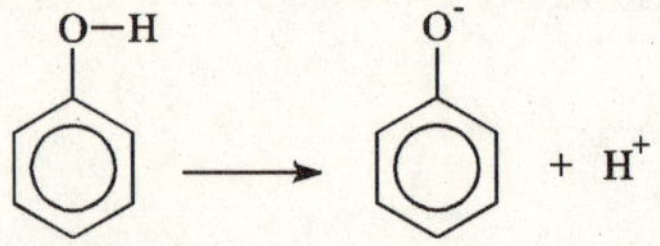

O–H → O^- + H^+

The way in which carboxylic acid groups and phenolic groups are bound into the complicated organic acids can be represented by

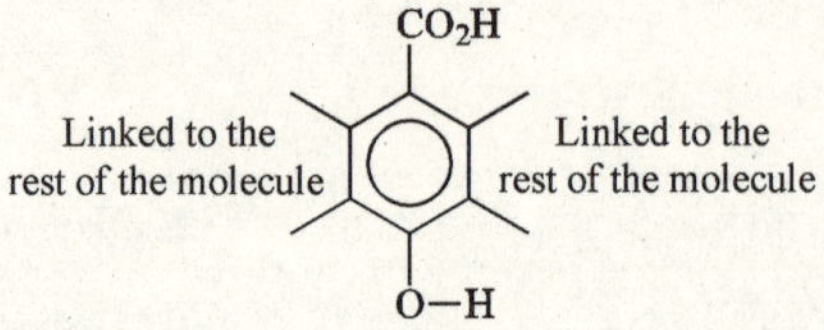

CO_2H

Linked to the rest of the molecule

Linked to the rest of the molecule

O–H

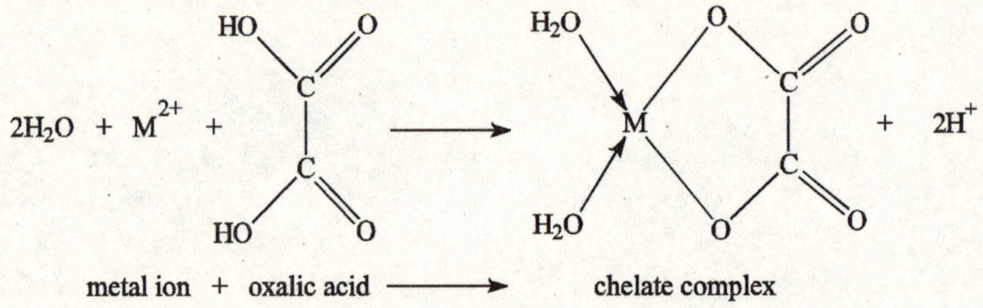

$2H_2O + M^{2+}$ + HO, O, C, C, HO, O → H_2O, H_2O, M, O, O, C, C, O, O + $2H^+$

metal ion + oxalic acid → chelate complex

Figure 8.1 Formation of a chelate complex (*see* Chapter 7) with a five-member ring.

soluble either after redox reactions have taken place or by chelate formation with organic acids. Again, they are easily removed in solution. Stable metal-organic chelates result from the formation of five-membered or six-membered ring systems by, for example, oxalic, citric or gluconic acid complexes (*see* Fig. 8.1 for an example).

8.2 Work of the Main Biological Agents

8.2.1 Work of bacteria

These microorganisms (*see* Box 8.2) have been active in the Earth's surface and near-surface environments for perhaps 3100 million years. They contribute to mineral alteration in a number of ways.

Box 8.2 Bacteria

These are unicellular or multicellular microscopic organisms, found widely in both aerobic environments, in which gaseous oxygen is present, and anaerobic environments, in which gaseous oxygen is absent. They may be *heterotrophic* (i.e. feed off organic molecules), *autotrophic* (obtain their nutrient from simple inorganic compounds) or *mixotrophic* (utilise both nutrient sources). They release carbon dioxide and carboxylic acids (*see* Box 8.1) and may be involved in a number of chemical processes (especially oxidation) that affect mineral stability.

Some important types for weathering purposes are the soil-dwelling *actinomycetes*, whose cells are arranged in rods or filaments; *nitrobacteria*, which convert ammonium ions to nitrates and nitrites (*see* Box 8.3); and *metallogenium* bacteria, which may oxidise manganese to produce 'desert varnish'.

By releasing carbon dioxide

Vast quantities of carbon dioxide are emitted by microorganisms, including bacteria, as part of their life processes. It has been estimated that the respiration of microorganisms may raise the proportion of carbon dioxide in soil air to 0.5–20 per cent. The atmospheric average is about 0.003 per cent. The resulting weakly-acid (pH 5–6) soil solutions enhance chemical weathering processes.

By engaging in the nitrification process (*see* Box 8.3)

This process is part of a larger chemical sequence called the *nitrogen cycle*. As part of this cycle, nitrogen is taken up by plants from nitrates in the soil, then absorbed by grazing animals, subsequently returned to the soil in the form of dead organic matter and finally converted to plant nutrient. Bacteria are involved in all these stages, and it seems that their exudates contribute to mineral weathering, but most attention has been focused on the activity of nitrobacteria, which are involved in the nitrification process.

Nitrification is the general term for the conversion of ammonium ions, which are products of the bacterial processing of dead plant and animal matter to nitrites (NO_2^-; the process of nitr*it*ation) and nitrates (NO_3^-; nitr*a*tation). The importance of this process for weathering is that it involves the release of hydrogen ions (protons). This can be seen from the general equation for nitratation, i.e. nitrate production:

$$\underset{\text{ammonium ion}}{NH_4^+} + \underset{\text{oxygen}}{2O_2} \longrightarrow \underset{\text{nitrate}}{NO_3^-} + \underset{\text{hydrogen ion}}{2H^+} + \underset{\text{water}}{H_2O}$$

In more detail, it is clear that two hydrogen ions are produced for every one ammonium ion. Consequently the hydrogen ion concentration increases in the soil solution and acidification occurs.

Box 8.3 Some terms used in the description of nitrification

Ammonia, NH_3

A compound occurring as a gas. It arises from the decomposition of dead organic matter, and is highly soluble in water.

Ammonium ion, NH_4^+

A positively charged ion (a cation) which may arise from the protonation of ammonia:

$$NH_3 + H^+ \longrightarrow NH_4^+$$

Nitrate ion, NO_3^-, and nitrite ion, NO_2^-

Negatively charged ions (anions) which result, for example, from the chemical conversion of ammonium ions.

Nitrobacteria, or nitrifying bacteria

(Nitric) acid-producing bacteria that oxidise ammonium ions to nitrite and to nitrate under neutral to alkaline conditions. They include the genera *Nitrosomonas* and *Nitrobacter*.

Experiments have been carried out, e.g. with vermiculite (*see* Chapter 10) from which Cu^{2+}, Mg^{2+} and Al^{3+} may be weathered by the H^+ released by bacterial action from added NH_4^+.

In Fig. 8.2 the release of Mg^{2+} from vermiculite in the presence of NH_4^+ is compared under inoculated (bacteria present) and sterile (bacteria absent) conditions. The release of Mg^{2+} after 50 days is approximately twice as much from the inoculated mixture, thereby demonstrating the importance of the nitrification reaction.

By the oxidation of metals

Many bacteria are capable of oxidising metals (*see* Box 8.4) which may then be released from their containing structures and subsequently precipitated. In extreme cases, commercial amounts of metals may be released from their ores, while in other circumstances only moderate amounts may be released and deposited.

The former case is illustrated by the action of thiobacilli (*see* Box 8.4). These oxidise reduced sulphur compounds such as metal sulphides which contain S^{2-} (e.g. ZnS, sphalerite), or S_2^{2-} (e.g. FeS_2, pyrites) to water-soluble metal sulphates (compounds containing the SO_4^{2-} ion). This microbial alteration is most effective under acid conditions. At pH 2–4.5, *T. ferrooxidans* can

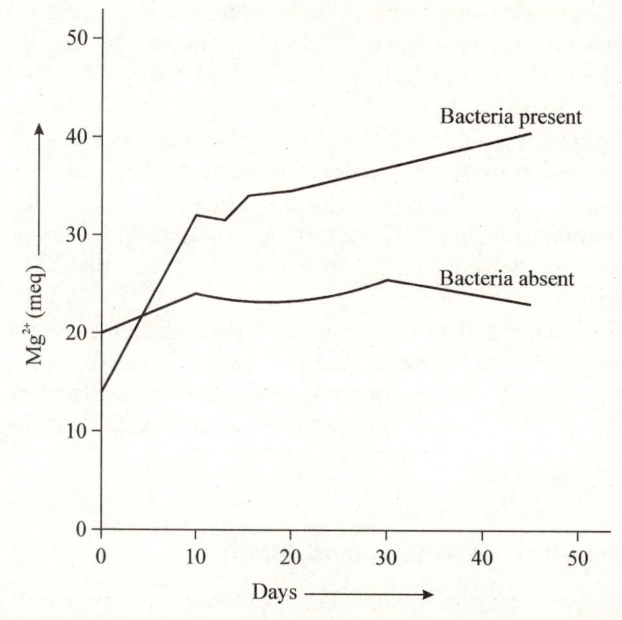

FIGURE 8.2 Release of magnesium from vermiculite. (Based on Simon-Sylvestre, G., Robert, M., Veneau, G. and Beaumont, A. 1991: Nitrification related to acidification and silicate weathering. In Berthelin, J. (ed.) *Diversity of environmental biogeochemistry*. Developments in Geochemistry 6. Amsterdam: Elsevier, 371–8.)

raise oxidation rates by factors of 10^5–10^6 compared to non-biological mechanisms.

The impressive ability of certain bacteria to 'leach' metal from ore has been used commercially. During the eighteenth century, copper was released bacterially from its ore in the mines of Rio Tinto, Spain. In 1971 about 15 per cent of the copper production of the United States was produced by bacterial 'leaching'.

Box 8.4 Metal-oxidising bacteria

The most effective of the metal-oxidising bacteria are the *thiobacilli T. thiooxidans* and *T. ferrooxidans*. Note that the prefix 'thio' means the replacement of oxygen by sulphur. They are termed *chemolithotrophic*. This means that they obtain their energy from the enzymatic oxidation of inorganic compounds or elements. Enzymes are proteins which accelerate chemical processes of metabolism. Perhaps of more interest in weathering are *mixotrophic* bacteria which use organic and/or inorganic nutrient sources. Manganese-oxidising types such as *metallogenium* bacteria may be important in the formation of desert varnish.

Other metal-oxidising bacteria have been used to explain the origin of '*desert varnish*'. This is a dark coating dominated by manganese and is found on rock surfaces in arid and semi-arid areas. The idea is based on the action of mixotrophic bacteria (*see* Box 8.2) for which these regions provide optimum conditions. Such bacteria are believed to oxidise the manganese in deposits on exposed rock surfaces. The process is most effective during sporadic moist periods, when it involves the cementing of clay fragments (micelles) as part of a precipitating manganese sheath. Such fragments help to shield the bacteria from harsh environmental conditions.

However, this view has not been universally accepted. A study of desert varnish on a plateau surface south of the High Atlas, Morocco, using both SEM (scanning electron microscope) and EPMA (electron microprobe) techniques has failed to reveal any trace of biological activity (*see* section 8.2.3).

8.2.2 The work of lichens (and fungi)

Lichens are compound plants (*see* Box 8.5) which are capable of carrying out both biophysical and biochemical weathering. The immediate evidence for this is provided by the zoned nature of the rock in the vicinity of a lichen (Fig. 8.3). This typically shows an upper region of rock fragments and a lower zone where the hyphae penetrate the solid rock. However, care is needed in interpreting field evidence. It is possible, for example, that rock breakdown may have been achieved before lichen colonisation.

Box 8.5 Lichens, algae and fungi

A *lichen* is a composite organism whose body (or *thallus*) consists of a fungus and an alga growing together in harmony. The fungus usually makes up the larger part of the thallus, with the algal cells contained in it. A useful subdivision of rock-colonising (i.e. saxicolous) lichens is into *endolithic* (growing within rock, e.g. in fractures) and *epilithic* (growing on the surface) varieties.

An *alga* (pl. *algae*) is a small, oxygen-producing green plant that varies in size from a single cell upwards and has not developed roots, stem or leaves. The blue-green algae, called *cyanobacteria*, have been important as oxygen producers over geological time.

A *fungus* (pl. *fungi*) is a plant which lacks chlorophyll and so obtains its nutrient from other plants or animals. Complex acids are released in exchange. The nutrient-absorbing and transmitting part of a fungus is the *mycelium*, made up of a complex web of fine threads, or *hyphae*, which penetrate rock surfaces or are buried in soil.

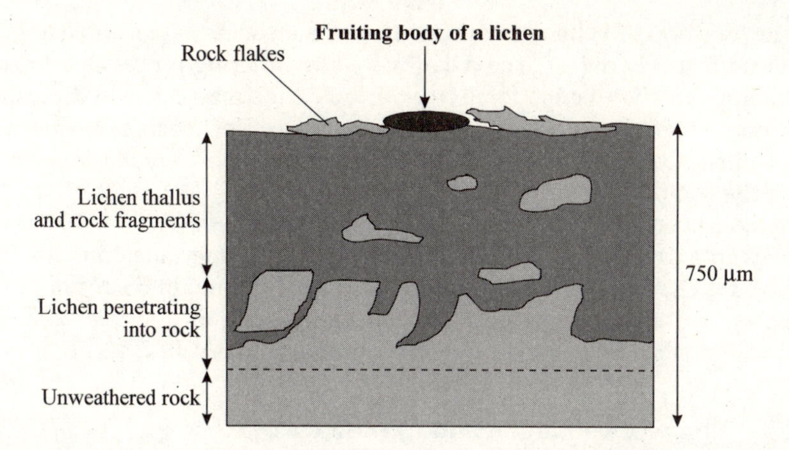

FIGURE 8.3 Typical banding of lichen and rock material in the surface weathered zone of a boulder. (Based on McCarroll, D. and Viles, H. 1995: Rock weathering by the lichen *Lecidea auriculata* in an arctic alpine environment. *Earth Surface Processes and Landforms* **20**, 199–206.)

Biophysical weathering

Mechanical breakdown of rock by lichens is brought about in two ways.

- **By the penetration of hyphae**. These grow along existing microcracks and between grains at a depth of between 1 and 3 mm and exceptionally up to 12 mm. Such growth may generate significant tensile stresses (see section 3.5) which are likely to be in excess of the tensile strength of most rocks.
- **By the expansion of thalli and hyphae due to water absorption**. Most lichens can increase their water content by 150–300 per cent under suitable conditions. Some whose thalli contain a high proportion of gelatine may show increases of up to 3900 per cent. The associated physical expansion may develop damaging levels of tensile stress.

Biochemical weathering

As part of their life processes, lichens emit organic acids which may be very effective in mineral alteration. For example, studies in Hawaii on lichen-colonised basalt have shown the development of a mature 'weathering front' on rock only 190 years old.

For a more detailed appreciation of the process it is necessary to recognise that the effectiveness of an organic acid depends on its solubility in water. It turns out that oxalic acid, probably produced by the fungal component, is soluble to the extent of $143\,g\,l^{-1}$. Protons are therefore provided and complexing may occur. Various oxalate compounds, e.g. magnesium oxalate dihydrate, or *glushkinsite*, have also been regarded as effective agents. However, the low solubility of oxalates means that their concentration in solution is low, which limits their ability to form complexes in weathering processes.

Other products of lichen growth, such as 'lichen acids', are soluble to levels between 5 and 57 $mg\,l^{-1}$. The details of the weathering process are unclear, but it appears that the H^+ ion in the various acids may cause hydrolysis. In addition, lichen acids act as agents of chelation. This process includes the mechanisms by which various acids combine with metallic ions in the rock, the compounds being absorbed by the lichens. Work on the Magaliesberg quartzite, near Pretoria, South Africa, suggests the importance of lichens in extracting compounds from rocks. The quartzite contains 98.08 per cent SiO_2; the same compound makes up a dominant 58.29 per cent of the chemical compounds found in lichens growing on the quartzite. This removal of silica from rocks, a kind of weathering, is called *desilication*.

8.2.3 The work of algae and cyanobacteria

These organisms may contribute to both the physical and chemical breakdown of rocks and minerals in a range of environments, including hot deserts and the Antarctic. The rock-dwelling types (lithophytic algae) are of greatest interest from a weathering point of view.

Biophysical weathering

The ability of certain algae to fracture rock has been studied on rock outcrops (nunataks) rising above the Juneau Icefield in Alaska. Here two species of green algae were found in microcracks formed by dilation (*see* section 6.5) and running parallel to the surface of exposed granite outcrops. These microcracks defined flakes between 2 and 4 mm thick, but of highly variable area, and which were easily dislodged under wet conditions. They could not be removed manually, however, after several days of hot weather. It was suggested that the polymer sheath, or *mucilage*, surrounding algal cells would take up a large quantity of water during wet periods, thus generating an expansionary force sufficient to lift already-weakened flakes away from the rock. The expansion may be as much as 20 times the volume of the dry state. An approximately similar process has been noted in Antarctica.

Biochemical weathering

Algae also appear capable of contributing to the dislocation and precipitation of metals at rock surfaces. Studies of desert varnish (see also section 8.2.1), a dark-brown to black coating typically less than 50 μm thick and found on a range of rock types in semi-arid environments, have shown, in Nevada and California, the apparent presence of manganese-rich *microstromatolites*. These are very small structures that may be produced by the action of blue-green algae. It has been suggested that manganese-oxidising cyanobacteria remove, by oxidation, manganese from wind-blown dust. This is then precipitated almost immediately, under episodically damp conditions, to form the varnish (*see* section 8.2.1 for a critical observation on the 'biological' hypothesis).

8.2.4 The work of plant roots

Remember that the upper zone of the regolith, dominated by plant roots, is referred to as the *rhizosphere*. Here roots may bring about rock and mineral breakdown by physical and chemical methods.

Biophysical processes

A commonly expressed idea is that the growth of plant roots in existing rock fractures may exert enough tensile stress for crack extension. A study of the mechanics involved should, however, persuade you that this process is unlikely. The measured axial growth pressure for many plants is about 3 MPa, which may fracture some weak rocks. Effective tensile stress is, however, set up by radial pressure, which is only about one-third to one-quarter of the axial amount. In addition, the applied force is likely to be even lower. The maximum force a root can exert is the product of its largest vertical cross-sectional area and its radial pressure. This area is likely to be much less than the area of the potential fracture surface, so a weak force is distributed over a large area. Mechanically, then, the traditional idea is unconvincing.

Biochemical processes

A plant root is a complex microsystem that both emits and absorbs substances as part of its life processes (*see* Fig. 8.4). The emitted materials include, from the tip to the main root, organic acids of low relative molecular mass (RMM), electrons (giving reduction) and protons (*see* Fig. 8.4(a)). These substances actively contribute to mineral decay mechanisms, among which one of the most important is the complexing process. This results in the release and mobilisation of aluminium and iron in chelate-like compounds (*see* Chapter 7). The complexing system also discourages crystallisation, leading to the formation of ferrihydrite (hydrated ferric oxide) and allophane-like minerals (allophane is an aluminosilicate gel). Several of the released (or available) substances are taken up by the root as part of the plant nutrient (*see* Fig. 8.4(b)). Calcium is particularly important. Studies

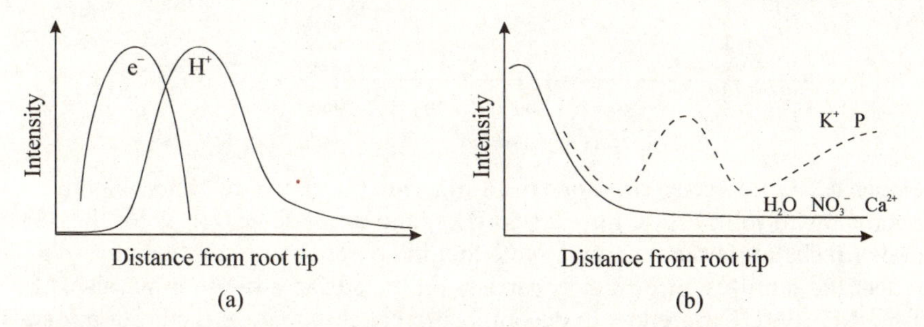

FIGURE 8.4 A schematic view of particles that are emitted (a) and substances absorbed (b), together with their locations along a root.

have shown that, for some calcic (calcium-rich) soils, more than 25 per cent of carbonate passes through the roots and may appear as calcite crystals occupying uniformly sized structures that were once root cells.

Roots and their surroundings are often associated with other organisms that enhance weathering. The rhizosphere typically shows an increase in the number of bacteria when compared with the rest of the regolith. In addition, a root may form a *mycorrhiza*, or close association with a fungus. Indeed, some plants, e.g. some species of *Pinus*, seem unable to develop normally if their mycorrhizal fungi are missing.

The rhizosphere is perhaps the most effective microsystem for weathering.

8.2.5 The work of decaying plant and animal matter

While many organic acids are excreted by living organisms, an important group is derived from the decay of soil organic matter. An advanced stage of such decay is represented by humus, which is a complex mixture of amorphous and colloidal substances, modified from the original plant and animal tissues or produced by soil organisms. The decay process, which is essentially oxidation, yields large amounts of carbon dioxide, reflecting the fact

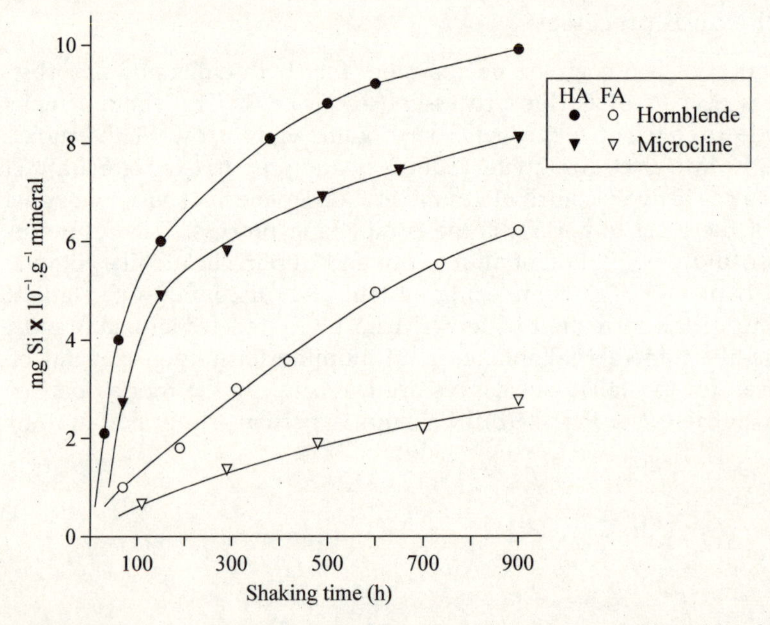

Figure 8.5 Silicon concentrations resulting from the exposure of hornblende and microcline to solutions of humic acid (HA) and of fulvic acid (FA) for the durations (in shaking hours) shown. Note that the parabolic dissolution curves may reflect the build-up of diffusion barriers on the surfaces of the minerals. (After Tan, K.H. 1991: Differences in decomposition of carbonate and silicate minerals as influenced by humic acid. In Berthelin, J. (ed.) *Diversity of environmental biogeochemistry*. Developments in Geochemistry 6. Amsterdam: Elsevier, 341–8.)

that carbon makes up over 50 per cent of the organic matter. Two organic acids, humic and fulvic, are also produced.

These agents belong to the class of acids that operate through proton and complexation reactions. They are both important chelating agents, especially fulvic acid (*see* Box 8.1). Typically a stable fulvic acid chelate is formed with Al^{3+} and Fe^{3+} or with aluminium and iron hydroxy ions, although the way in which mineral surfaces are attacked is not clear. Because such chelates are soluble, they are removed from the weathering environment. Elsewhere, relatively stable calcium-humic acid may be formed.

Experimental work has thrown light on the relative effectiveness of humic and fulvic acids in releasing elements from minerals. Finely ground samples of calcite, hornblende and microcline were shaken in fulvic and humic acids for over 800 h, and the amount of Si, Al, Fe, K and Ca released were determined. The results showed that humic acid was more effective than fulvic acid in releasing elements from the silicates used, while the two acids were equally effective in the dissolution of calcium from calcite. Figure 8.5 shows representative results.

READINGS FOR FURTHER STUDY

Unless you have studied biology you may find the ideas in this chapter rather unfamiliar. Beginners can look up basic biological ideas in Curtis, N. 1985: *Longman Illustrated Dictionary of Biology*, Longman Group.

The chapter in Prof. Yatsu's book (see below) gives an overview of biological weathering, but it is detailed and you should be prepared to skip sections.

Drever, J.I. 1994: The effect of land plants on weathering rates of silicate minerals. *Geochimica et Cosmochimica Acta*, **58**, 2325–2332. Provides a useful link with the material of Chapter 9.

Eckhardt, F.E.W. 1985: Solubilization, transport, and deposition of mineral cations by micro-organisms - efficient rock weathering agents. In Drever, J.I. (ed.) *The chemistry of weathering*. Dordrecht: D. Reidel, 161–73. Reviews the contribution of bacteria, algae, fungi and lichens to weathering processes.

Hall, K. and Otte, W. 1990: A note on biological weathering of nunataks of the Juneau Icefield, Alaska. *Permafrost and Periglacial Processes* **1**, 189–96. A discussion of the ways in which algae may break down granitic rocks.

Smith, B.J. and Whalley, W.B. 1988: A note on the characteristics and possible origins of desert varnishes from southeast Morocco. *Earth Surface Processes and Landforms* **13**, 251–8. Considers organic and inorganic origins for a rock varnish.

Viles, H.A. (ed.) 1988: *Biogeomorphology*. Oxford: Basil Blackwell. The contribution of biological studies to landform problems.

Yatsu, E. 1988: Weathering by organisms. In *The nature of weathering: an introduction*. Tokyo: Sozosha, 285–396. A detailed review of a large number of academic papers on the processes of biological weathering. Use selectively!

9

INTENSITY AND RATE OF WEATHERING

The terms 'intensity' and 'rate' refer to two distinct concepts. The *intensity*, or degree, of weathering is the amount of alteration from its original state shown by solid rock or unconsolidated sediment as a result of the various decompositional processes. Essentially, weathering intensity refers to the degree of decomposition **at a point in time**. *Rate*, on the other hand, refers ideally to the amount of change **per unit time**, although in practice it refers to generalised change. The calculation of rate requires a knowledge of the time period during which alteration has taken place, and so it is not often achieved outside the laboratory. The two ideas may be linked in that a high degree of weathering intensity may imply a rapid alteration rate. However, care is needed as a high intensity may also be achieved by modest rates acting over a long time. This may be the case in the humid tropics.

In this chapter the term 'measured rate' will be used where a specific time period is involved, but the term 'rate' may be used for generalised observations where quantitative measurements are not available.

9.1 THE IDEA OF EQUILIBRIUM, OR STEADY STATE, IN WEATHERING

A natural system, such as a beach or river channel, is said to be in equilibrium when its characteristics, reflecting a balance between external processes and material properties, stay unchanged over time. In reality, processes especially change, and so a true equilibrium is rarely maintained. The same argument applies to the regolith. The factors controlling weathering are constantly changing, so it is not possible for a true equilibrium to exist at the Earth's surface. However, the dominance of stable minerals in some weathering systems suggests that a sort of quasi-equilibrium may be achieved. This is especially the case in tectonically stable parts of the humid tropics, where high temperatures, ample water and long undisturbed periods of weathering (>100 000 years) result in mineral assemblages which show little change and are more or less in equilibrium with their surroundings.

Such assemblages are rich in one or both of the more stable components Al_2O_3 and Fe_2O_3, which are end products of the *ferrallitic* weathering trend. Water (H_2O) and silica (SiO_2) are also present. The dominant phases are quartz (SiO_2), goethite ($FeO(OH)$), kaolinite ($Al_2(OH)_4Si_2O_5$) or gibbsite ($Al(OH)_3$). Quartz and gibbsite, however, are not in theory expected to coexist in a stable equilibrium. If they were to occur together, kaolinite should form until one or the other is used up. There is evidence that gibbsite is produced in quartz-free systems, while kaolinite forms when quartz is present.

Outside the humid tropics, such quasi-equilibria do not occur. The *podzolic* weathering trend (*see* Box 10.2) may be taken as an example. This is characteristic of humid, cool-temperate climates. The E horizon still contains weatherable minerals, especially feldspars. This suggests that equilibrium has not been reached, a view that is consistent with the relatively short time (~10 000 years, the time since the last ice sheets melted) that the podzolic trend has been active.

9.2 Controls of intensity

The intensity of weathering, or the extent to which the regolith approaches a state of quasi-equilibrium, is determined by the interaction of a set of factors. Essentially, these factors affect the rate and nature of the weathering process. Where a rapid rate is encouraged there may be a high degree of weathering, but slow rates may achieve a similar end product if sufficient time is available.

The controlling factors fall into two categories: *intrinsic* and *extrinsic*. The intrinsic factors include the properties of the parent material (*see* Chapter 3), such as the nature and organisation of pores and fractures, and mineralogical composition. The extrinsic factors include environmental temperature, fluid chemistry and hydrodynamics.

9.2.1 Intrinsic factors

Pores and fractures

These physical properties guide the entry of weathering fluids and so influence the intensity of weathering. They operate firstly at the scale of the *rock outcrop*, where the pattern and intensity of weathering is typically a function of fracture (joint, bedding plane or fault) density. They act secondly at the *mineral scale*, where alteration follows cracks, fissures, cleavage planes and other surfaces of weakness.

Mineralogy

Different minerals might be expected to show differing responses to weathering processes, and several attempts have been made to arrange minerals in the order of their stability. A natural method of doing this might be to arrange them by the frequency of their occurrence in sedimentary rocks of increasing age. This approach would imply an averaging out of the great variety of environments that have occurred in the geological past. Minerals would essentially be arranged in the order of their persistence, which may be interpreted as degree of stability. This approach was proposed by Pettijohn (1941; *see* Table 9.1).

This approach, and those of various other investigators, suggested general agreement about the relative stability of the feldspars and quartz, and little

TABLE 9.1 Pettijohn's sequence of mineral persistence based on mineral frequency in sedimentary rocks of increasing age

	Mineral	Persistence
	Anatase	(–3)*
	Muscovite	(–2)
	Rutile	(–1)
Stability decreasing ↓	Zircon	(1)
	Tourmaline	(2)
	Monazite	(3)
	Garnet	(4)
	Biotite	(5)
	Apatite	(6)
	Ilmenite	(7)
	Magnetite	(8)
	Staurolite	(9)
	Kyanite	(10)
	Epidote	(11)
	Hornblende	(12)
	Andalusite	(13)
	Topaz	(14)
	Sphene	(15)
	Zoisite	(16)
	Augite	(17)
	Silimanite	(18)
	Hypersthene	(19)
	Diopside	(20)
	Actinolite	(21)
	Olivine	(22)

[a] Negative numbers indicate a tendency to formation rather than disappearance during long periods of burial.

Note that quartz was not considered

Source: Pettijohn, F.J. 1941. Persistence of heavy minerals and geologic age. *Journal of Geology* **49**, 610–625.

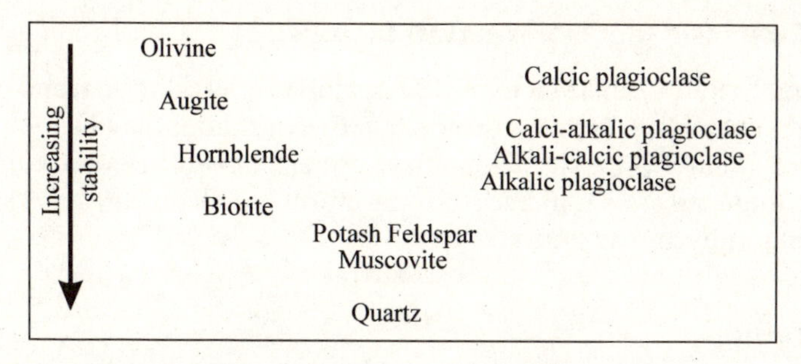

FIGURE 9.1 Goldich's sequence of mineral susceptibility. It is similar to Bowen's sequence of minerals crystallising from a melt. (*Source*: Goldich, S.S. 1938. A study of rock weathering. *Journal of Geology*, **46**, 17–58.)

dispute about that of the more common ferromagnesian minerals. Perhaps the most generally accepted expression of the traditional view was the contribution by Goldich (1938). He analysed chemical and mineralogical changes in two igneous and two metamorphic rocks from the New England and Central Lowland Provinces of the United States, and from his results derived a mineral stability series for common igneous rock-forming minerals (Fig. 9.1). His sequence of increasingly stable minerals is similar to N.L. Bowen's sequence of mineral crystallisation from a melt.

This similarity may be interpreted as suggesting that the stability of a mineral species *increases* as the difference between the temperature of formation and that of the weathering environment *decreases*. You should remember that both temperature and pressure may affect the position of a chemical equilibrium, and so if the conditions of formation and of final state are close together, the material will be closer to equilibrium with its surroundings.

Goldich's sequence of mineral susceptibility has been the dominant theory in weathering studies for the subsequent 50 years. Criticisms have, however, emerged fairly recently. It has been pointed out that there is no convincing mechanism to account for Goldich's sequence, and that data are emerging that conflict with his order of mineral susceptibility. In particular, studies of Hawaiian igneous rocks in various weathering environments have demonstrated the importance of environmental conditions. When ample organic acids were present, derived from bacteria, algae, fungi and lichens, the weathering sequence matched the Goldich model. Olivine weathered first, followed by clinopyroxene (augite) and then (calci–alkalic) plagioclase. Elsewhere, however, under near-sterile conditions, the reverse sequence was found. The clinopyroxenes and plagioclase feldspars weathered *before* adjacent olivines. It was argued that most of the early work on weathering sequences relied on data from samples taken from environments where organic acids were present, and that under dry (xeric) conditions the classic findings did not apply. Therefore the importance of environmental conditions appears quite clear.

9.2.2 Extrinsic (environmental) factors

Traditionally the extrinsic factors include climatic, vegetational and geomorphological controls. However, these act *indirectly*, and it may be more logical to replace them by those factors that impact *immediately* on weathering intensity and rate. These include temperature, the chemistry of weathering solutions and hydrodynamics.

Temperature

The main importance of this factor is its control on reaction rate (*see* Chapter 7). Generally, for every 10°C rise in temperature, the velocity of a chemical reaction increases by a factor of two. Geographically there are considerable variations in average temperature, and so from polar to tropical regions, for example, reaction rate increases by about 10 times. The data in Table 9.2 suggest, at a simple level, the effect of temperature on measured weathering rates.

A rise in temperature may cause a weathering process to take place different to that operating at a lower temperature. For example, the solubility of minerals is affected by changes in temperature and the nature and number of ions in aqueous solution. A particular illustration of this is the precipitation of gibbsite ($Al(OH)_3$), which will tend to occur at higher temperatures and where the activity of silicic acid (H_4SiO_4) is low.

TABLE 9.2 Data suggesting the effects of temperature on weathering rates in different environments

Locality	Type of rock or mineral	Rate (μm per 1000 years)
Baffin Island (Canada)	Hornblende	0.1–0.5
Montana	Hornblende	<1
Western USA	Basalts	10–20
Southern Alps (New Zealand)	Arenites (sandstones)	1 000
Central Europe	Granites	6 000–13 000
Chad	Granite, gneiss	13 500
New Caledonia	Acid rocks	10 000–16 000
New Caledonia	Ultrabasic/basic rock	29 000–47 000
Ivory Coast	Granites	14 000
Ivory Coast	Granites	5 000–50 000
West Africa	Volcanics	10 000–50 000

Note that the rates are for particular sites, and that geological, geomorphological and biotic influences are contributory factors.

Source: Colman, S.M. and Dethier, D.P. 1986: An overview of rates of chemical weathering. In Colman, S.M. and Dethier, D.P. (eds) *Rates of chemical weathering of rocks and minerals*. London: Academic Press, 1–18.

Chemistry of weathering solutions

The effectiveness of weathering solutions is largely determined by their pH, or proton concentration, as well as by temperature. The pH of rainfall is typically in the range 5.5–6.2. You may remember that the pH of pure water is 7. Rainwater has a lower (i.e. more acidic) pH owing to the presence of dissolved carbon dioxide, which is an acidic gas. The pH may be affected locally by factors such as the presence of dust or of aerosols from the sea. Human activities also provide important contaminants. For example, acid rain may be produced by emissions of sulphur dioxide (SO_2) from fossil fuel-fired power stations (*see* Box 6.1). Precipitation usually comes into contact with vegetation and picks up further compounds. The net effect is to decrease the pH by 1 or 2 units.

The pH is further lowered as water passes through the upper layers of the soil. This is because the soil air usually holds between five and 22 (rarely to 400) times the atmospheric content of carbon dioxide. In addition, organic acids (bioacids) may be important sources of protons (*see* Box 8.1). A variety of acids (acetic, formic, lactic, oxalic, malic and citric) has been recorded in podzols of the Vosges mountains, in concentrations between 3 and $92\,\mu g\,l^{-1}$. Lower down the profile the pH tends towards neutrality. Table 9.3 gives some examples of soil solutions.

While the effectiveness of the weathering solution is largely driven by pH, it is subsequently modified by contact time with the earth material. The style and rate of water movement (hydrodynamics) are therefore very important.

Hydrodynamics

Clearly, for weathering to take place, water must circulate through rock. The significance of water migration can be seen by contrasting the weathering behaviour of igneous and sedimentary rocks. The former might be expected to weather more rapidly than the latter, as their constituent minerals are generally less stable. However, soils tend to form more readily on sedimentary rocks than on other types, and the reason is their open structure which allows rapid water circulation. Conversely, where igneous rocks are permeable (e.g. non-welded products of explosive volcanic behaviour) they weather readily, while slightly permeable sedimentary rocks (e.g. the Carboniferous limestones of England and Wales) are typically resistant.

The pattern of water migration varies between two broad classes of earth material. In **solid rock** water will exploit existing passages at all scales. At the smallest scale, that of the individual mineral, water enters fractures and cleavage planes, and varying styles and degrees of alteration will develop (*see* Fig. 9.2), determined by the patterns of weakness, temperature, and chemical characteristics of the water.

At a larger scale, that of joints, bedding planes and faults, the degrees of weathering will be strongly affected by the density and pattern of these fractures. This point is developed in Chapter 11.

TABLE 9.3 Some examples of the composition of weathering solutions in different environments

Locality	Soil type	Horizon	pH	Cation content ($\mu mol\ l^{-1}$)					
				Ca	Mg	Na	K	Si	Al
Cold									
Washington State, USA	Podzol	A	–	180	80	40	40		150
		Bh/Bs	–	20	80	10	10		270
Galicia, NW Spain	Cambisol[a] (on granite)	A	3.6–6.6	12–300	21–987	204–7739	8–569	18–295	0–2
Temperate	Cambisol (on serpentinite[b])	A	6.4–7.5	68–360	113–1040	278–1209	2–100	64–311	0–2
Tropical dry									
Chad	Vertisol[c]	A	7.5	1740	860	1335	685	443	–
Tropical wet									
Queensland	Ferralsol[d]	A	4.1–7.9	10–640	50–860	700–860	100–790	–	–
		B	4.6–6.6	5–100	5–100	130–400	10–180	–	–

[a] A *cambisol* has a moderately weathered B horizon; [b] a *serpentinite* is a rock once rich in olivine that has been altered to serpentine; [c] a *vertisol* has a high content of swelling clays; [d] a *ferralsol* is high in iron and aluminium.

Source: Based on a table in Macia, F. and Chesworth, W. 1992: Weathering in humid regions with emphasis on igneous rocks and their metamorphic equivalents. In Martini, I.P. and Chesworth, W.P. (eds) *Weathering, soils and palaeosols*. Developments in Earth Surface Processes 2. Amsterdam: Elsevier, 283–306.

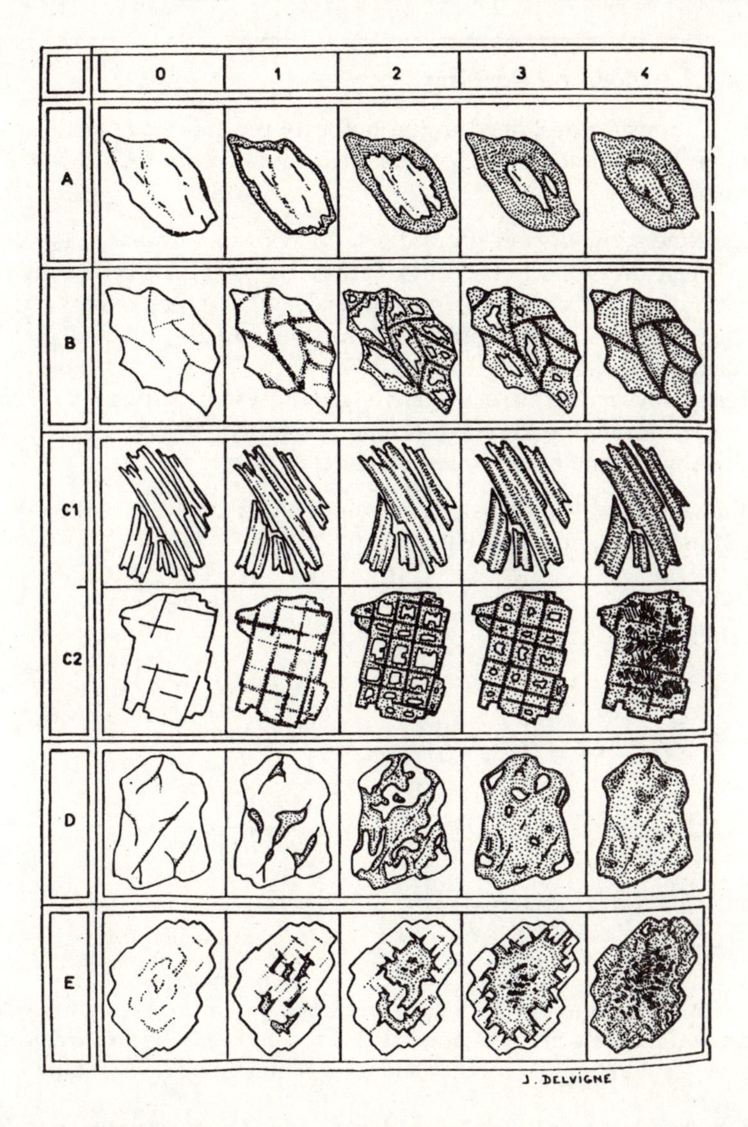

FIGURE 9.2 Alteration sequences, showing the different patterns and degrees of alteration in a range of minerals of varying structural type. Key: A, Pellicular alteration of olivine to iddingsite; B, irregular linear alteration of olivine to nontronite; C_1, parallel linear alteration of muscovite to kaolinite; C_2, cross linear alteration to K-feldspar into amorphous material and feldspar; D, dotted alteration of olivine into nontronite; E, complex alteration of K-feldspar into amorphous material and kaolinite. 0, Fresh or nearly fresh, <2.5 per cent of mineral altered; 1, slightly altered, 2.5–25 per cent of mineral altered; 2, moderately altered, 25–75 per cent of mineral altered; 3, strongly altered, 75–97.5 per cent of mineral altered; 4, completely altered, >97.5 per cent of mineral altered.
(*Source*: Stoops, G. *et al.* 1979: Guidelines for the description of mineral alterations in soil micromorphology. *Pedologie* **29**, 121–35.)

Box 9.1 More on flushing

The relationship described in the text between the flushing rate and the rate of dissolution can be illustrated by reference to two types of terrain.

- *Limestone terrains*. Here groundwater may be close to saturation with calcite ($CaCO_3$), i.e. the ratio between dissolved calcite and the saturation concentration (maximum level) is greater than 0.5. In this case calcite dissolution is probably greatly affected by the flushing rate.
- *Feldspar-dominated terrains*. Here groundwater is usually highly undersaturated with reference to the dissolving feldspars. The dissolution rate is controlled by feldspar reactivity.

Within a generally undersaturated solution there are different types of dissolution behaviour (*see* Fig. 9.3).

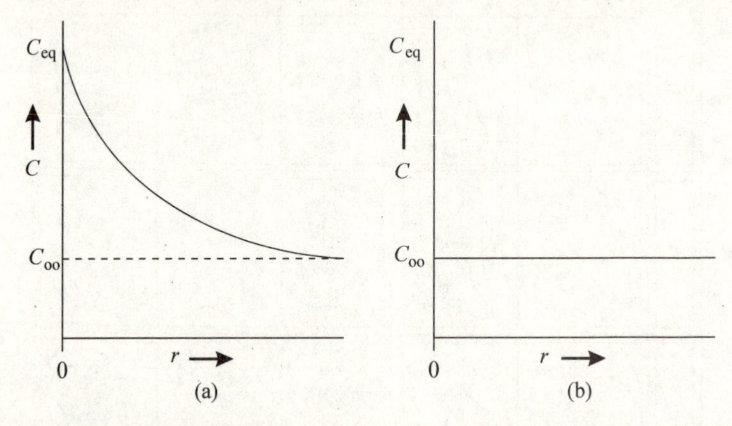

Figure 9.3 Concentration (C) versus distance (r) from a crystal surface for two rate-controlling processes: (a) transport control and (b) surface reaction control. C_{eq} = saturation concentration, C_{oo} = concentration in solution.

- *Transport-controlled dissolution* occurs when ions are released so rapidly that they build up against a crystal. The dissolution rate is controlled by the transport of such ions into the adjacent under-saturated solution.
- *Surface reaction-controlled dissolution*. In this case the concentration at the crystal surface is the same as in the surrounding solution, and increased flow has no effect on the dissolution rate.

(Based on material in Berner, R.A. 1978: Rate control of mineral dissolution under earth surface conditions. *American Journal of Science* **278**, 1235–52.)

The pattern is different in the **regolith**. For humid regions, three generalised pathways of water flow are found.

- Groundwater flow.
- Shallow, saturated flow in soil or saprolite. This is subsurface storm flow, and typically occurs at the junction between A and B horizons where permeability is reduced.
- Saturation overland flow, which occurs where subsurface water returns to the land surface, or where precipitation falls onto saturated areas, usually near surface channels.

The organised flow of water through consolidated rocks and regoliths contributes in two main ways to weathering rate and intensity.

First, relatively fresh water is brought into contact with minerals and weathering may be accelerated. Solution contact times with the regolith are generally short, however, being between a few hours and a few days for subsurface storm flow, and shorter for saturation overland flow.

Second, the products of weathering are carried away in solution by 'flushing'. There is clearly some sort of relationship between this process and the rate of dissolution. Theoretical investigations suggest the following.

- As the flushing rate is increased, the concentration of solutes decreases and the rate of dissolution increases. A point is reached at which the maximum rate of dissolution is independent of the flushing rate.
- If flow is sluggish, the concentration of solutes increases and the rate of dissolution decreases. Ultimately saturation is reached and the rate of dissolution is controlled only by the rate of flushing (*see* Box 9.1).

Generally, flushing speeds up the dissolution of minerals in water-saturated rocks only up to a limiting rate beyond which dissolution is controlled by the solubility of minerals.

9.3 GLOBAL EXPRESSION OF INTENSITY

The factors driving weathering, especially temperature, the nature of the weathering solutions, and the hydrodynamics of the weathering material, vary globally, and we might expect that broad weathering zones would develop as a consequence.

Figure 9.4 is a well-known expression of these global variations. It shows the important climatic factors of precipitation, temperature and evaporation, together with the depths and nature of the weathering mantle, along a transect from the north polar regions to the hot humid tropics. The great weathering zones are clearly shown.

9.3.1 Polar region

The polar region is a province of shallow weathering and minimal chemical activity, the major product being allophane (*see* Chapter 10). This pattern is a

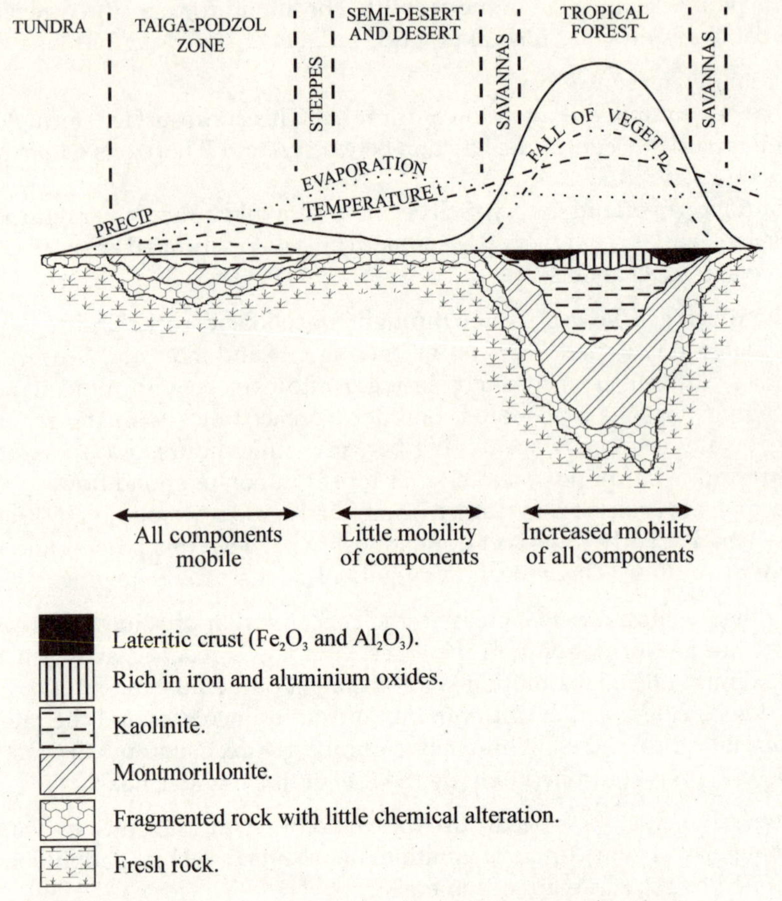

FIGURE 9.4 Depth of weathering and associated weathering products as a function of environmental factors along a latitudinal transect. (Based on a diagram in Strakhov, N.M. 1967: *Principles of lithogenesis*. Edinburgh: Oliver and Boyd.)

result of low temperatures and slight precipitation. These factors diminish biomass production, rate of breakdown of organic matter, leaching potential, proton availability and formation of chelating agents (*see* Chapter 7).

9.3.2 'Taiga–podzol zone'

This is a region of much more active weathering. The cool to temperate climate favours the accumulation of humus at a more rapid rate than the microbial population can break it down. Chelation (*see* Chapter 7) is important, and encourages the mobilisation of aluminium and iron as organo-metallic complexes. The combination of chelation and leaching favours horizon development, with an upper silicon-rich E horizon contrasting with a lower Al–Fe–organics-rich B horizon (*see* Box 10.2).

However, the presence of a mixture of primary and secondary phases, including allophane, in the B horizon, which has been described as a 'salad bowl', suggests that this environment is far from being a weathering equilibrium.

9.3.3 'Steppes' and 'semidesert and desert' zones

In these regions the temperature rises, precipitation falls and evaporation increases towards low latitudes. Chemical weathering is reduced, in spite of high temperatures, by lack of water; horizon development is suppressed; and the tendency is for salts, e.g. gypsum ($CaSO_4 \cdot 2H_2O$), to accumulate. Weathering intensity is therefore slight.

9.3.4 'Savanna' and 'tropic forest' zones

These are provinces where a combination of high temperatures and increasing precipitation favour the growth of vegetation. This is especially the case in the tropical forest zone, where plant matter decays rather than accumulates, and where the thickest and most developed regoliths occur.

The end stage of weathering is made up of the components SiO_2, Al_2O_3, Fe_2O_3 and H_2O. The predominant phases are quartz, goethite, kaolinite (dominant in well-drained sites) and gibbsite (found where drainage is very rapid). Laterite and other crusts may form where water availability is lower.

The weathering trend is termed *ferrallitic* (i.e. iron 'ferr' and aluminium 'all' rich) because of the tendency of all rocks to change towards compositions rich in one or both of the components Fe_2O_3 and Al_2O_3.

Humid tropical weathering comes close to reaching phase equilibrium.

9.4 Assessment of weathering intensity

The intensity or degree of alteration shown by a regolith can be described by methods varying from the verbally descriptive to the scientifically objective. Typically, the length of time over which weathering has occurred is not considered, and so measurements of intensity usually have little bearing on weathering rate, the determination of which is discussed in section 9.5.

9.4.1 Verbally descriptive approach

A simple appraisal of the intensity of weathering is particularly useful in developing countries, where a rapid reconnaissance for engineering purposes may be required. Such an appraisal may be based on observation

TABLE 9.4 Scheme of weathering grades for engineering purposes, worked out for granitic rocks

Term	Grade	Description of weathering intensity
Fresh	I	No visible sign of rock material weathering; perhaps slight discoloration of major discontinuity surfaces
Slightly weathered	II	Discoloration indicates weathering of rock material and discontinuity surfaces. Some of the rock material may be discolored by weathering; yet it is not noticeably weakened
Moderately weathered	III	Less than half of the rock material is decomposed or fragmented. Discolored but unweakened rock is present either as a continuous framework or as corestones
Highly weathered	IV	More than half of the rock material is decomposed or disintegrated. Fresh or discolored or weakened rock is present either as a discontinuous framework or as corestones within the soil
Completely weathered	V	All rock material is decomposed and/or disintegrated to soil. The original mass structure and material fabric are still largely intact
Residual soil	VI	All rock material is converted to soil. The mass structure and material fabric are destroyed. There is a large change of volume, but the soil has not been significantly transported. Can be divided into an upper A horizon of eluviated soil and a lower B horizon of illuviated soil

Source: Based on British Standard BS 5930: 1981. Code of practice for site investigations.

leading to verbal description. An example, used for engineering description, is given in Table 9.4.

9.4.2 Description based on the use of tools or instruments

As well as visual description, weathering intensity may be assessed instrumentally. The equipment employed ranges from an implement such as a boot or a hammer used to make a simple assessment of strength, to rather more sophisticated devices such as the Schmidt hammer and seismic equipment.

A simple approach involves a five-point scale of friability (the extent to which the rock can be broken up), given in Table 9.5. This scale is based on the assumption that there is a positive correlation between simply measured rock strength and the degree of weathering. A more advanced technique involves the Schmidt hammer. This approach avoids the problem of operator bias. This instrument is used to measure changes in the mechanical strength of a rock. Such changes may be due to weathering, and so are a measure of the degree of alteration.

TABLE 9.5 A five-point scale of friability

1. Fresh: a hammer tends to bounce off rock
2. Easily broken with a hammer
3. The rock can be broken by a kick (with boots on) but not by hand
4. The rock can be broken apart in the hands, but does not disintegrate in water
5. Soft rock that disintegrates when immersed in water

Source: Ollier, C.D. 1965: Some features of granite weathering in Australia. *Zeitschrift für Geomorphologie* **9**, 285–304.

Some results of Schmidt hammer tests are shown in Fig. 9.5, which demonstrates the unusual case of weathering actually leading to rock strengthening.

A more elaborate instrumental approach has been to measure the velocity of compression waves generated by hammer blows against, for example, granitic rock. The velocity reduces as microfracture density increases; the magnitude of the reduction is taken as a measure of weathering intensity.

9.4.3 Measures of intensity based on chemical indices

As weathering proceeds, and the more mobile elements and compounds are removed, the chemical composition of a regolith will change. There are two broad approaches to measuring the intensity of such change. First, the degree of alteration can be assessed in absolute terms by comparing the constitution of the parent material, where this is known, with that of the weathered formation. The second approach is to work out the ratio between

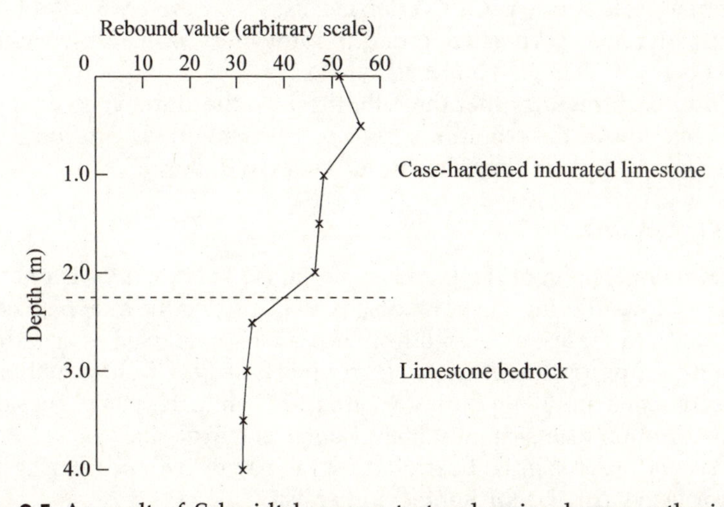

FIGURE 9.5 A result of Schmidt hammer tests, showing how weathering may occasionally lead to rock hardening. 'Case-hardened indurated' means that the limestone surface has been hardened by solution and precipitation of more stable crystalline structures. (*Source*: Day, M.J. and Goudie, A.S. 1977: Field assessment of rock hardness using the Schmidt test hammer. *British Geomorphological Research Group Technical Bulletin* **18**, 19–29.)

a highly resistant mineral, such as quartz, and others which are more easily removed. This ratio is an index of the intensity of weathering.

'Absolute' methods

The principle behind the 'absolute' approach is that a quantifiable relationship can be established between unweathered and weathered material. Two absolute methods have been used.

Isovolumetric method
This is perhaps the simplest way of measuring the loss of rock components through weathering. It is based on the assumption that there has been no change in volume between rock in the unweathered and weathered states (hence *isovolumetric*). Evidence for this assumption would be the preservation of the original petrographic textures or geological structures. This tends only to be the case when the intensity of weathering is quite low. The method has therefore been applied to the 'arènes' (coarse granitic saprolites) of Western Europe, which do not show the intense alteration of tropical regoliths.

The method involves a comparison of the content of, say, an oxide in unweathered and weathered samples. The difference, usually a loss in the weathered sample, is a measure of the amount of weathering that has taken place. An illustration of the isovolumetric method is given in Box 9.2.

'Benchmark mineral' method
In order to relate the composition of the regolith to that of the unweathered parent material, and so calculate loss, a resistant mineral must be selected as a benchmark. Alumina (Al_2O_3) is often used as it is relatively insoluble at pH values down to 5.5, such as are found in rainwater. Even after weathering, alumina is retained in clay minerals.

The approach requires that the ratio between the alumina content of the parent rock and of the regolith is used to calculate the loss of other compounds. An example of the calculation is shown in Box 9.3.

Relative methods

When the composition of the parent material is not known, or is uncertain, the degree of weathering can be assessed by calculating the ratio between the more stable and the less stable oxides. The result is expressed as an index. At this point the approaches differ. One group is based on the assumption that alumina remains immobile during weathering, while the other relies on the same assumption for resistant minerals such as tourmaline.

The second approach is illustrated by the use of two weathering ratios (WR), for heavy (h) and for light (l) minerals:

$$\mathrm{WR(h)} = \frac{\text{zircon} + \text{tourmaline}}{\text{amphiboles} + \text{pyroxenes}}$$

$$\mathrm{WR(l)} = \frac{\text{quartz}}{\text{feldspars}}$$

Box 9.2 Illustration of the isovolumetric method of determining oxide loss by weathering

The method has been applied to an arène developed on the Turiz granite in the Cavado river basin between Braga and the coast, Portugal. The figures for selected oxides are given in Table 9.6, and their derivation is described below.

TABLE 9.6 Oxide content by weight of parent rock and arène, and absolute and percentage loss due to weathering

	Oxides percentage	Oxide weight g $(100\,cm^3)^{-1}$			Isovolumetric balance			
	RS	RS	AEC		Weight variation			
			base (B)	top (T)	g $(100\,cm^3)^{-1}$		(%)	
	Turiz	1	2	3				
	granite	$d_1 = 2.68$	$d_2 = 1.71$	$d_3 = 1.47$	Δ2	Δ3	2	3
	(1)	(2)	(3)	(4)	(5)	(6)	(7)	(8)
SiO_2	66.1	177.1	104.8	88.2	−72.3	−88.9	−41	−50
Al_2O_3	15.3	41.0	30.3	25.9	−10.7	−15.1	−26	−37
Fe_2O_3	4.6	12.3	10.8	9.8	−1.5	−2.5	−12	−20
TiO_2	0.94	2.5	2.2	1.9	−0.3	−0.6	−12	−24
MnO	0.07	0.2	0.2	0.1	0.0	−0.1	0	−50
MgO	1.47	3.9	4.2	3.3	+0.3	−0.6	+8	−15
CaO	2.60	7.0	1.4	0.6	−5.6	−6.4	−80	−91
Na_2O	2.91	7.8	1.4	0.7	−6.4	−7.1	−82	−91
K_2O	4.86	13.0	8.5	7.7	−4.5	−5.3	−35	−41
Other	0.44	1.2	7.3	6.5	+6.1	+5.3	+508	+442
Total	99.29	266.1	170.9	144.7	−95.2	−121.4	−36	−46

Note: RS = parent rock; *d* = density; AEC = arène.

1. The oxide content of the parent rock (column 1) is multiplied by 2.68 to give an oxide weight g $(100\,cm^3)^{-1}$ (column 2).
2. The oxide weights are worked out for the base (density 1.71) and top (density 1.47) of the arène (columns 3 and 4).
3. Simple subtraction gives the weight changes for the base and top horizons (columns 5 and 6). The changes are expressed as percentages (columns 7 and 8).

Source: Based on Braga, M.A.S., Lopes Nunes, J.E., Paquet, H. and Millot, G. 1990: Climatic zonality of coarse granitic saprolites ('arènes') in Atlantic Europe from Scandinavia to Portugal. In Farmer, V.C. and Tardy, Y. (eds) *Proceedings of the 9th International Clay Conference*, Strasbourg, 1989. *Science Géologique, Mémoire* **85**, 99–108.

The latter ratio has been used to investigate the intensity of granite weathering on Dartmoor, southwest England.

Most approaches have, however, been based on the assumption that alumina remains immobile during weathering, and Table 9.7 shows some of the ratios that have been presented. You should note that the method

Box 9.3 Calculation of weathering losses from a quartz–feldspar–biotite gneiss

Oxide	Adjusted		Corrected figure for each oxide, ratio 14.61/18.40	Loss from parent rock (2)–(4)	Percentage change (5)/(2) × 100
	Fresh rock	Saprolite			
(1)	(2)	(3)	(4)	(5)	(6)
SiO_2	71.48	70.51	55.99	−15.49	−22
Al_2O_3	14.61	18.40	14.61	0	0
Fe_2O_3	0.69	1.55	1.23	+0.54	+78
FeO	1.64	0.22	0.17	−1.47	−90
MgO	0.77	0.21	0.17	−0.60	−78
CaO	2.08	0.10	0.08	−2.00	−96
Na_2O	3.84	0.09	0.07	−3.77	−98
K_2O	3.92	2.48	1.97	−1.95	−50
H_2O	0.32	5.90	4.68	+4.36	+1360
Others	0.70	0.54	0.43	−0.27	−39
Total	100.00	100.00	79.40	−20.60	

Source: Data from Krauskopf, K.B. 1967: *Introduction to geochemistry*. New York: McGraw-Hill.

Explanation

Columns 2 and 3 show the chemical constitution, in percentages, of fresh rock and saprolite. The alumina (Al_2O_3) content in the saprolite has increased from 14.61 to 18.40 because other oxides have been lost. So the corrected *total* for oxides is in the ratio 14.61/18.40, i.e. from 100 to 79.4. The corrected figures for *each* oxide can be calculated (column 4) from the ratio. For example:

$$SiO_2 \text{ loss} = \frac{70.51\,(\text{column } 3) \times 14.61}{18.40} = 55.99$$

The actual loss is calculated by subtracting the adjusted oxide figure from that for the parent rock, e.g. for SiO_2 it is $71.48 - 55.99 = 15.49$. The percentage loss is calculated by dividing the actual loss by the parent rock content and multiplying by 100, e.g. for SiO_2 it is

$$\frac{15.49}{71.48} \times 100 \simeq 22\%$$

used to write down the amounts of various elements is based on the idea that they occur as oxides. This is satisfactory for the calculation, but you should recognise that actually compounds such as sodium oxide (Na_2O) would react rapidly with atmospheric moisture to form sodium hydroxide:

$$Na_2O + H_2O \longrightarrow 2NaOH$$

TABLE 9.7 Some weathering indices in common use and which (apart from WI) rely on the assumption that aluminium remains immobile during weathering

Chemical index of weathering:

$$CIW = [Al_2O_3/(Al_2O_3 + CaO + Na_2O)] \times 100$$

Chemical index of alteration:

$$CIA = [Al_2O_3/(Al_2O_3 + CaO + Na_2O + K_2O)] \times 100$$

Weathering index:

$$WI = [(2Na_2O/0.35 + (MgO/0.9) + 2K_2O/0.25) + (CaO/0.7)] \times 100$$

Weathering index of Reiche (1943), modified by Vogel (1975):

$$MWPI = [(Na_2O + K_2O + CaO + MgO)/(Na_2O + K_2O + CaO + MgO + SiO_2 + Al_2O_3 + Fe_2O_3)] \times 100$$

Vogt ratio (Vogt, 1927; Roaldset, 1972):

$$V = (Al_2O_3 + K_2O)/(MgO + CaO + Na_2O)$$

Ruxton ratio (Ruxton, 1968):

$$R = SiO_2/Al_2O_3$$

Source: Chittleborough, D.J. 1991: Indices of weathering for soils and paleosols formed on silicate rocks. *Australian Journal of Earth Sciences* **38**, 115–20.

Sodium hydroxide is itself reactive. The sodium ions in weathered material are actually present as ions attached, for example, to particles of clay minerals or within the silicates (*see* Chapter 10).

A simple illustration is given by the 'Ruxton ratio' (silica-to-alumina mole ratio). This is based on the assumption that silica loss correlates significantly with total element loss. This ratio was recommended for use in most free-draining environments in humid climates, especially on acid and intermediate rocks that weather to illite, but not for strongly seasonal climates where silica may indeed migrate and/or be precipitated.

The basic assumption that alumina remains stable during weathering may not, however, be justified. Recently it has been found that the movement of water through a weathering profile may displace layer-lattice silicates of submicrometre size which contain aluminium. These silicates are moved from the surface to lower horizons. As a consequence the upper horizons would appear, from commonly used indices, to be less weathered than the lower horizons! This conclusion may be inconsistent with other evidence of weathering intensity, such as the bleached quartz-rich and feldspar-deficient nature of the upper horizons compared with the lower ones.

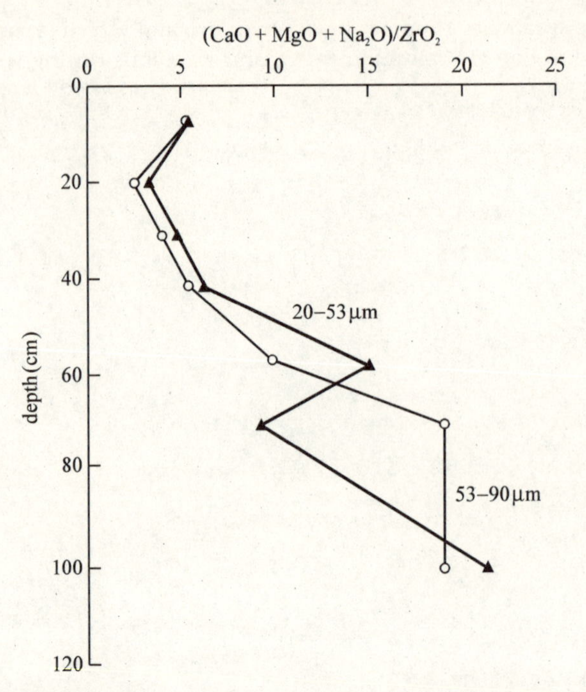

FIGURE 9.6 The weathering ratios at various depths for two fractions of a soil developed on a feldspathic sandstone near Adelaide, South Australia. (*Source*: Chittleborough, D.J. 1991: Indices of weathering for soils and paleosols formed on silicate rocks. *Australian Journal of Earth Sciences* **38**, 115–20.)

A weathering ratio (WR) based on resistant heavy minerals in the 20–90 μm fraction ($1\,\mu m = 10^{-6}$ metres) may therefore be better than the traditional approaches, since this would exclude the submicrometre layer-lattice silicates discussed above:

$$WR = [(CaO + MgO + Na_2O)/ZrO_2]$$

Here the resistant mineral is zircon (ZrO_2), but other stable minerals such as tourmaline, rutile (TiO_2) and xenotime may also be used. The fine sand fraction is employed, since this is where these minerals are normally concentrated. The weathering ratios plotted in Fig. 9.6 show how weathering intensity varies with depth.

9.4.4 Measures of intensity based on the characteristics of individual mineral grains

As weathering proceeds, individual minerals are likely to show alteration, and the degree of alteration may be seen as a measure of weathering intensity. We describe two methods using this approach.

TABLE 9.8 Surface microtextural features of heavy mineral grains used for determining weathering class (for further explanation *see* Box 9.4)

Surface feature no.	Assigned feature	Point value	Maximum value assessment
Denotes 'freshness'			
1	High relief	+1	+50
2	Clean/smooth fracture or cleavage faces	+1	+50
3	Arc-shaped/parallel/semiparallel steps	+1	+50
4	Sharp edges and angular grains	+1	+50
5	Conchoidal fractures/breakage blocks/v-shaped pits	+1	+50
Denotes weathering			
6	Solution	−1	−50
7	Scaling or surface roughness	−1	−50
8	Oriented or random etch pits	−1	−50
9	Subdued edges and rounded grains	−1	−50
10	Hairline cracks	−1	−50

Source: Tejan-Kella, M.S., Chittleborough, D.J. and Fitzpatrick, R.W. 1991: Weathering assessment of heavy minerals in age sequences of Australian sandy soils. *Journal of the Soil Science Society of America* **55**, 427–38.

Method based on microtextural characteristics of resistant heavy minerals

Microscopic (SEM; *see* Box 3.6) examination of the surface of a heavy mineral (i.e. density $>2.96\,g\,cm^{-3}$) typically reveals a range of features that may be called 'microtexture' (*see* Table 9.8). These features may be divided into those suggesting a fresh unweathered state, and those indicative of weathering.

The nature and intensity of the features may be quantified (*see* Box 9.4) and a degree of weathering can be determined (*see* Table 9.9). As a consequence, an order of mineral resistance to weathering can be established, and for regoliths in three different coastal regions in eastern and southern Australia the following sequence has been proposed:

order of decreasing resistance ⟶

zircon > sillimanite ≥ spinel ≥ rutile > garnet > epidote > monazite

(*Source*: Tejan-Kella, M.S., Chittleborough, D.J. and Fitzpatrick, R.W. 1991: Weathering assessment of heavy minerals in age sequences of Australian sandy soils. *Journal of the Soil Science Society of America* **55**, 427–38.)

Box 9.4 Assessment of weathering class, based on microtextural study

Each grain in a population is examined by SEM and allocated a positive or negative point value (PV) according to the presence of 'fresh' or 'weathering' features (*see* Table 9.8). A point value score (PVS) up to ±5 is derived from the sum of the point values for that grain. The mean of 30 PVs for each grain type is the mean PVS (MPVS). Minerals are assigned to a weathering class according to the MPVS.

However, the *extent* and *magnitude* of microtextural features must also be taken into account. The positive and negative PVs are turned into value assessments (VAs) (*see* Table 9.9) by multiplying the value of the *degree* of development by the value of the *area* of development. The maximum VA is ±50, based on a degree-of-development value of ±5, and an area-of-development value of 10. The VA score (VAS) is derived from the sum of individual VAs. A maximum VAS range of ±250 is possible. The mean VAS (MVAS) for the 10 classes of surface feature on 30 grains is used to define degree of weathering, and minerals are assigned accordingly (*see* Table 9.9).

Method based on the development of 'cockscomb' morphology in hornblende grains

This approach is similar to that discussed above in that microtextural character is used to determine the degree of weathering. In this case the surface property is cleavage-controlled etching at the edge of hornblende grains, giving a 'cockscomb' appearance which can be measured. The method has been chiefly used to measure weathering rate, and so is more fully discussed in section 9.5.2.

TABLE 9.9 A weathering classification for heavy mineral grains

Class no.	Weathering classification	Mean point-value score	Mean value-assessment score range
1	Highly weathered	−4, −5	−150 to −250
2	Weathered	−2, −3	−50 to −149
3	Somewhat weathered	1, 0, −1	49 to −49
4	Slightly weathered	3, 2	149 to 50
5	Essentially unweathered	5, 4	250 to 150

Source: Tejan-Kella, M.S., Chittleborough, D.J. and Fitzpatrick, R.W. 1991: Weathering assessment of heavy minerals in age sequences of Australian sandy soils. *Journal of the Soil Science Society of America* **55**, 427–38.

9.5 WEATHERING RATES

The rate at which rocks and minerals break down is called the 'weathering rate'. It may be defined more precisely as a measured deterioration in an earth material during a specified time period.

A knowledge of weathering rates has both academic and practical importance. Such information may help to date episodes in the geomorphological evolution of an area, and may help to clarify environmental conditions during the Quaternary (the last 1.6 million years). It has also strong practical applications. Work in Sweden has suggested that the weathering rate, and so proton consumption, of certain acid soils is too slow to compensate for the loss of base cations through leaching and harvesting. A consequence is increasing acidity and diminishing fertility. Weathering rates may also be used to date landsurfaces and so improve our knowledge of the frequency of hazardous events such as landslides.

Various units, typically of volume or distance per unit time, are used to describe rates, and different expressions of rate are possible (*see* Box 9.5). Weathering rates may be uniform (or linear) over time, as is probably the case with the short-term export of ions from a drainage basin. More commonly, however, the rate varies with time and so is non-linear, as may be the case with the growth of a weathering rind on a rock fragment.

A major problem in the determination of weathering rates from field evidence is the difficulty of arriving at an accurate numerical (absolute) date for an event in the history of a weathering substance. This event may, for example, be a glaciation which deposited an unweathered sediment. If the date of such a glaciation is known, and the degree of weathering of the upper horizons of the sediment is measured, then a weathering rate can be calculated. Increased accuracy is possible if several related events can be dated. This approach is difficult to apply in practice because of the shortage of datable events.

9.6 CALCULATION OF WEATHERING RATES

There are two broad approaches to the determination of weathering rates. The first is to carry out experiments in the laboratory. This approach allows factors such as time, mineralogy and the surface area of the experimental substance to be controlled, but there are important drawbacks. These include the simplification required by laboratory experiments, and the limited time during which the procedures are carried out. In addition, the active agents often used in experimental solutions, e.g. fulvic acid and acetylacetone, are usually at much higher concentrations than occur in the environment.

The alternative approach is to study naturally weathered materials, or the chemistry of waters draining catchment areas. The problems with this approach derive from the complexity of the real world, and especially the

Box 9.5 Some units in which weathering rates are measured

Before weathering rates are discussed, we need to define some units which are used to describe *quantities*, i.e. *units of loss*.

Some units for measuring alteration by ionic loss:

- $meq\, m^{-2}$ (milliequivalents per square metre);
- $mol\, ha^{-1}$ (or cm^{-2}) (moles per hectare, or per square centimetre);
- $mol\, g^{-1}$ (moles per gram).

Typical dimensional units for expressing loss/change:

- μm (micrometers, micro = 10^{-6});
- mm (millimetres, milli = 10^{-3}).

Having defined units of loss, we can now define *rates*. Where alteration is uniform (or linear) over time, rates may be expressed as alteration in unit time:

- $meq\, m^{-2}\, a^{-1}$ (milliequivalents per square metre per year);
- $mol\, ha^{-1}\, (cm^{-2})\, a^{-1}\, (s^{-1})$ (moles per hectare (per square centimetre) per year (per second));
- $\mu m\, ka^{-1}$ (micrometers per 1000 years);
- $mol\, g^{-1}\, h^{-1}$ (moles per gram per hour);
- $t\, km^{-2}\, a^{-1}$ (tonnes per square kilometre per year).

More commonly, the relationship between alteration and time is non-uniform (non-linear), and may be expressed as a regression equation. For example, the rate of accumulation of clay in the B horizon of an Antarctic soil has been expressed as

$$Y = a + (b \log X)$$

where y is clay present as a percentage, a and b are regression coefficients, and X is time, in years.

By entering some data the equation becomes

$$Y = -7.88 + (2.23 \log 10^6)$$

$$= 5.5\% \text{ clay accumulation in 1 million years}$$

variety of earth materials and of environmental conditions. It is extremely difficult to isolate factors for study. In addition, the shortage of dated events in recent Earth history creates problems in calculating weathering rates when naturally weathered materials are studied.

9.6.1 Calculation of rates from laboratory experiments

Chemical weathering

For chemical weathering certain key questions have been illuminated by laboratory investigations.

TABLE 9.10 Some mineral weathering rates determined by laboratory experiment

Mineral	Weathering rate ($mol\ cm^{-2}\ s^{-1}$)	Notes
Feldspars		
Plagioclase group		
Labradorite	1.58×10^{-16}	Release rate of Ca
	8.45×10^{-17}	Calculated from Si release rate
Albite	1.6×10^{-16}	
	3.12×10^{-16}	
	4.2×10^{-16}	
Oligoclase	$\sim 1 \times 10^{-16}$	
Alkali feldspar	7.33×10^{-16}	
Olivine	3.98×10^{-15}	Release rate of Mg from forsterite
	5.01×10^{-16}	Calculated from Si release rate
Garnet	1.1×10^{-15}	
Amphibole		
Tremolite	1.45×10^{-15}	
	2.0×10^{-16}	
Hornblende	1.82×10^{-16}	
	$(4.2–8.3) \times 10^{-16}$	

Source: Based on material in Velbel, M.A. 1993: Constancy of silicate-mineral weathering-rate ratios between natural and experimental weathering, etc. *Chemical Geology* **105**, 89–99.

Do weathering rates vary between minerals?

Much work has been done on the response of silicates, and especially feldspars, to low-temperature (i.e. environmental) weathering solutions in the laboratory. Even under such controlled conditions, different laboratories may produce findings that vary by a factor up to 30 for the same mineral. Some detailed weathering rates are given in Table 9.10. It is possible to make some generalisations from the data about comparative mineral resistance:

- olivine weathers 6–25 times faster than calcium plagioclase (labradorite);
- garnet weathers about four times faster than sodium plagioclase (albite);
- amphiboles and sodium plagioclases weather at about the same rate.

Earlier laboratory investigations had shown that calcium plagioclases weather more rapidly than sodium plagioclases.

These findings are consistent with the relative resistance of olivines, amphiboles and calcium and sodium plagioclase in the Goldich weathering sequence (*see* Fig. 9.1).

Is the rate of weathering affected by mineral structure?

At the microscale it appears that both the original and developing structure of a mineral will affect its weathering rate. Weathering occurs most rapidly at reactive sites such as crystal defects, dislocations, twin boundaries or

Table 9.11 Mineral properties and weathering rates

Minerals		Lattice coherence[a]	Diffusion avenues		Weathering rate	
Primary	Secondary		Initial	Later	Predicted	Observed
Olivine	Smectite	None	Sparse dislocations	Wide	Fast	Very fast
Pyroxene	2:1 layer	Close	?Exsolution lamellae	Small	Slow	Slow
Amphibole	2:1 layer	Close	?	?Small	Slow	?Slow
Muscovite	Smectite	Very close	Interlayer	Narrow interlayer	Very slow	Slow
Biotite	Vermiculite, goethite, etc.	Close	Interlayer	Moderate (wrinkles)	Medium	Medium
Feldspar	Clay	None	Twins, dislocation	Wide etch channels	Fast	Fast

[a] The term 'lattice coherence' refers to the ease with which the minerals fit together.
Source: Modified from Eggleton, R.A. 1986: The relation between crystal structure and silicate weathering rates. In S.M. Colman and D.P. Dethier (eds) *Rates of chemical weathering of rocks and minerals*. London: Academic Press, 21–40.

microfractures. As weathering develops, its rate depends on whether these weaknesses become enlarged, or blocked with secondary products. In the former case the weathering rate is maintained, while in the latter it decreases. Work using TEM (transmission electron microscopy, allowing resolution of ~3 Å (*see* Box 3.7)) has shown that diffusion paths are easily blocked if the structures of the weathering mineral and of its secondary products are similar.

These ideas can be illustrated by reference to two minerals. The structural units of the clays that result from olivine breakdown are too large to occupy diffusion avenues in the mineral, and so alteration is rapid. On the other hand, TEM investigations of pyroxene hydration show that the first-formed weathering product, a talc-like layer silicate, has lattice dimensions that almost exactly fit those of the primary mineral. The initial weathering rate is therefore low, but accelerates later as different alteration products evolve.

Table 9.11 shows the predicted and observed weathering rates for a number of minerals. These rates are derived from the lattice relationship between the minerals and their alteration products, and are linked to diffusion avenue development.

Does the rate of weathering decrease over time?
Early experimental studies supported the idea that ion release from a weathering mineral, notably feldspar, slows down with time, i.e. the relationship is non-linear. This effect was explained as resulting from the development either of a leached layer or of a precipitated layer at the mineral surface. Such a layer would slow down mineral alteration. However, later studies

of the surfaces of feldspar, amphibole, pyroxene and olivine grains have not revealed such a layer.

It now appears that the non-linear behaviour of feldspars observed in early experiments was due to preparation techniques. For example, grinding releases fine powder at a mineral surface, and this will react rapidly with weathering fluids until it is dissolved. A high apparent initial rate of mineral dissolution will fall away once these fine particles have been dissolved.

Is the rate of mineral weathering affected by the solution used?

The properties of a weathering solution will affect the rate of mineral alteration. Experiments have shown that the release rates of aluminium and silicon are clearly different at different pH values. It has also been observed that the presence of organic chelating agents (*see* section 8.1.3) in solution may affect the rate of mineral alteration. For example, a highly concentrated solution of fulvic acid (found in natural waters; *see* Box 8.1) increases the dissolution rate of chlorite. The position is not straightforward, however, as other work has shown that the dissolution rate of calcite may be decreased by organic chelation (*see* section 7.12).

Is the weathering rate proportional to mineral surface area?

It has seemed reasonable to assume that the weathering rate, or reaction rate, is proportional to the specific surface area ($m^2\,g^{-1}$) of a mineral, i.e.

$$\text{rate} = (\text{surface area})^{m}$$

where m is an exponential power function. If the rate were directly proportional to area, then m would be nearly 1. However, an uncertainty is introduced by the fact that weathering is not uniform but is selective, concentrated at crystal defects. In addition, experimental work on feldspars has shown that for grains smaller than 38 μm, exponents may vary between −0.02 and 0.18, i.e. weathering is not proportional to specific surface area.

It now appears that the weathering rate of feldspars varies according to the size of grains involved. A schematic plot (*see* Fig. 9.7) of rate against surface area shows two regions of contrasting activity.

- *Region I* shows the typical response of coarse grains (>38 μm) where defect spacing is less than the dimensions of individual grains. It follows that the number of defects exposed per unit surface area should stay the same over a range of grain sizes. Reaction rates will be proportional to a specific surface area, i.e. $m \simeq 1$.
- *Region II* behaviour applies when the dimensions of individual grains are similar to or less than crystal defect spacings. In this case any additional crushing of the mineral (i.e. reduction in size) will not expose defects at a rate equal to that of the production of a new surface area. Reaction rates will not increase as rapidly as the specific area, or perhaps not at all, i.e. $m \simeq 0$.

Reaction rate appears therefore to be a function of a critical particle size, itself dependent on defect spacing.

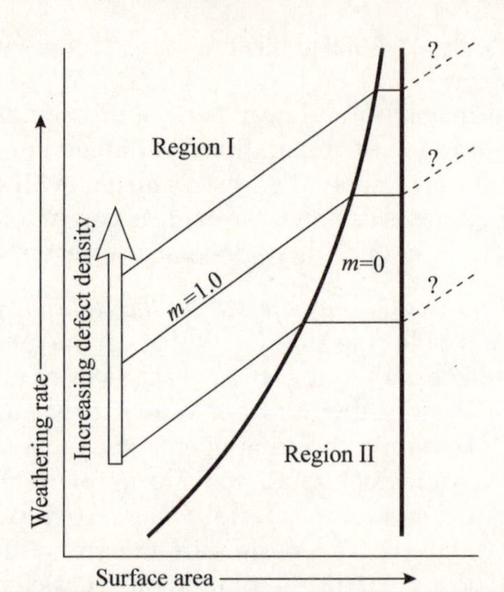

Figure 9.7 A schematic illustration of rates versus surface area plots expected from minerals having different defect densities. (*Source*: Holdren, G.R. and Speyer, P.M. 1985: Reaction rate–surface area relationships during the early stages of weathering – I. Initial observations. *Geochimica et Cosmochimica Acta* **49**, 675–81.)

Mechanical weathering (cf. Chapter 6)

The rocks used in laboratory experiments on the rates of physical weathering are usually much more complex than the relatively simple minerals used in dissolution experiments. Conclusions about breakdown rates are correspondingly more difficult to draw, but it does seem that rate is strongly related to rock type.

This may be illustrated by French work on frost shattering. Rocks in the form of 9 cm cubes were placed in basins which were then filled with water to a depth of 1–2 cm. A light freezing regime was used, with the air temperature falling to −8°C. This simulated the daily freezing regime characteristic of polar maritime regions. The results showed that the highest rates of breakdown occurred in chalks, marly limestones, molasse sandstone and soft argillaceous rocks. Fifty per cent of a particular chalk sample was reduced to particles of less than 50 μm after 500 freeze–thaw cycles. In comparison, dense, extremely fine-grained limestone produced less than 3 per cent fragments of less than 50 μm after 500 cycles, and unweathered Precambrian schists produced less than 2 per cent after 1000 cycles.

9.6.2 Calculation of rates from field-based studies

Several approaches have been used in attempts to work out weathering rates from field evidence. These approaches involve the study of

- the degree of alteration in weathered profiles;

- the chemical content of water draining from catchments;
- weathered rocks and minerals.

Weathered profiles

Over time the weathering of a soil or superficial deposit will bring about the release of certain elements and compounds which may then be removed by migrating water. The amount of loss tends to be greatest near the ground surface and to diminish with depth. This trend is illustrated in Fig. 9.8(a), which shows the depletion of certain cations in a forested watershed of the Adirondack Park region of New York State.

It might also be expected that easily weathered minerals such as hornblende would show a similar trend, and this is demonstrated in Fig. 9.8(b) for the same region. Conversely, more resistant minerals such as zircon ($ZrSiO_4$), quartz (SiO_2) and opaque minerals such as magnetite (Fe_3O_4) might be relatively more common in the upper parts of a profile, where less resistant types have been removed, than in the lower horizons, where weathering is less intense. This common trend is illustrated in Fig. 9.8(c), also for the Adirondack region.

The rate of element and compound loss can be worked out by relating the amount of loss to the age of the sediment. The former can be determined if both the original and present-day proportion of a substance can be measured by comparison with stable and so immobile reference substances. It is clear that this approach only provides information about long-term weathering rates.

Weathering rate studies based on these principles have been made on soils developed in glacial tills. One such study involved seven sites in Scandinavia where sandy tills ranged in age between 8800 yrs and 120 000 yrs BP (Before the Present). It was assumed that zirconium (Zr) in the mineral form zircon was the stable element, and that the original concentrations of the unstable elements Ca, Mg, K and Na were the same as those in the present C horizon. Clearly the concentration of zircon at higher horizons increases as the more easily weathered elements are removed (cf. Fig. 9.8(c)). Measurements were made of the elemental composition of the soil mineral material of size less than 2 mm at about 10 horizons (levels). The loss of elements could then be calculated (*see* Box 9.6).

The results showed that the annual losses of material since deglaciation were on average 2–20 $g\,m^{-2}$. Variations between sites were due to factors such as geochemistry and climate (especially temperature). These annual rates, which are essentially historical, should not be confused with present-day rates.

An important practical consequence of this type of investigation is that it can throw light on the balance between the rates at which fertility-related elements are contributed by weathering and lost by plant uptake and leaching. For example, a study in central Sweden showed negative magnesium balances for about 70 per cent of the area examined. This type of situation may have serious long-term consequences for soil fertility.

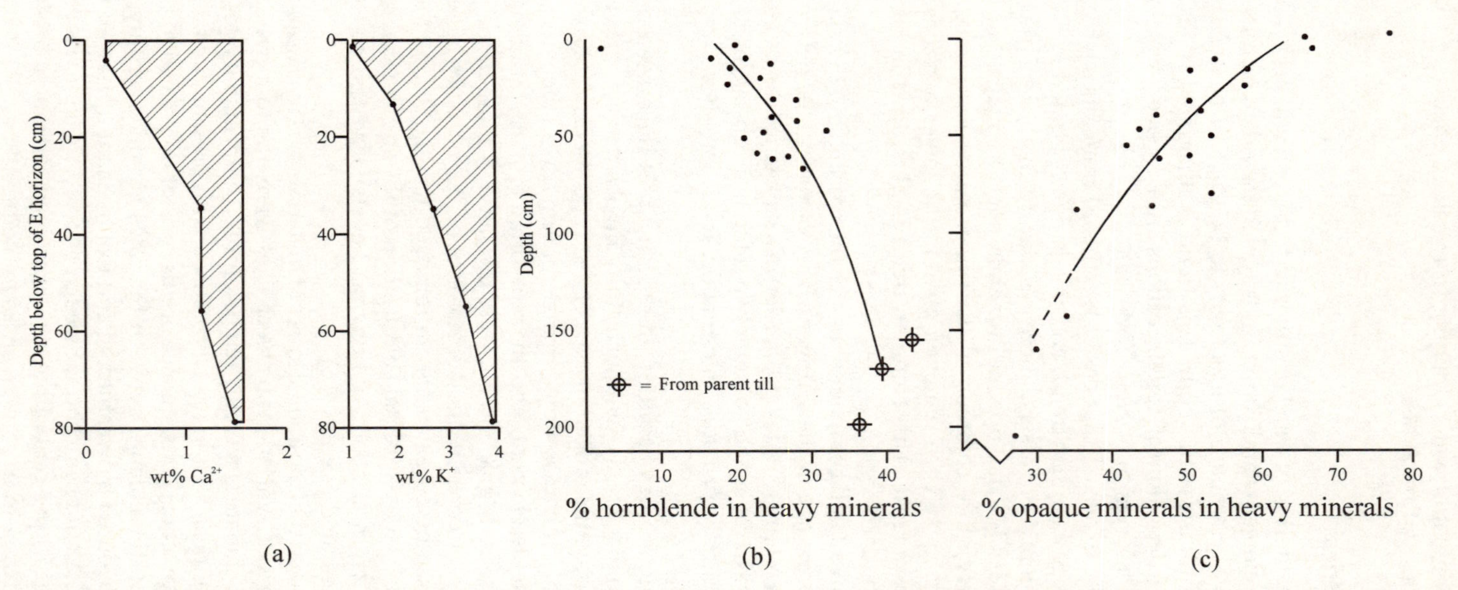

FIGURE 9.8 Variation of certain cations and minerals with depth in a forested watershed of New York State. (a) Examples of cation depletion curves, (b) trends in hornblende abundance with depth and (c) trends in abundance of resistant opaque minerals (ilmenite and magnetite) with depth. (*Source*: April, R., Newton, R. and Coles, L.T. 1986: Chemical weathering in two Adirondack watersheds: past and present-day rates. *Geological Society of America Bulletin* **97**, 1232–8.)

Box 9.6 Calculation of weathering loss

A knowledge of the present concentration and amounts of a resistant element, e.g. zirconium (Zr) which mainly occurs in the mineral zircon ($ZrSiO_4$), allows a calculation of loss based on the following equations:

$$W' = (100 \times Zr^W/Zr^C) - W^W$$

$$X' = (X^C \times Zr^W/Zr^C) - X^W$$

where

W' = loss of soil material <2 mm [$g\,m^{-2}$] in the weathered part, i.e. where Zr concentrations exceed those in the C horizon;

W^W = present amount of soil material <2 mm [$g\,m^{-2}$] in the weathered part;

Zr^W = present amount of Zr [$g\,m^{-2}$] in the weathered part;

Zr^C = concentration of Zr, as a percentage of soil material <2 mm in the weathered part;

X' = loss of element X (i.e. Ca, Mg, Na or K) [$g\,m^{-2}$];

X^C = concentration of the element X, as a percentage of material <2 mm in the C horizon;

X^W = present amount of element X [$g\,m^{-2}$] in the weathered part.

(*Source*: Based on material in Olsson, M. and Melkerud, P.A. 1991: Determination of weathering rates based on geochemical properties of the soil. *Geological Survey of Finland, Special Paper* **9**, 69–78.)

Catchment (geochemical mass balance) studies

In principle, the measurement of the ionic content of water draining a catchment can be used to work out the weathering rate (i.e. rate of ion release) of the rocks forming that catchment. In practice, however, the calculation is made uncertain because of various complications.

- *Ions contributed from the atmosphere and by biological processes must be subtracted from the gross content in drainage waters.* Chloride, sulphate, magnesium, sodium and other ions are present in precipitation, and calcium, sulphate, potassium and other cations may be deposited by 'dry deposition', especially in drier areas. There are few reliable measurements of these additions.
- *Vegetation, and soil organic matter, are reservoirs that contain large amounts of cations such as calcium, magnesium and potassium.* Investigators often assume that these reservoirs are in a steady state and so may be ignored

TABLE 9.12 Some weathering rates derived from catchment studies; note the variety of units used

Locality	Rate	Notes
Southern Blue Ridge, USA	$38\ mm\ ka^{-1}$	Rate of saprolite formation
Pacific North West	$30\ mm\ ka^{-1}$	Saprolite formation
Virginia	$7\ mm\ ka^{-1}$	Saprolite formation
Idaho catchments, USA	$26.7\ t\ km^{-2}\ a^{-1}$	Plagioclase weathering
Northeast Scotland	$6–8\ t\ km^{-2}\ a^{-1}$	Gneiss and gabbro
Pacific North West	$34\ t\ km^{-2}\ a^{-1}$	$160\,000\ km^2$
Czechoslovakia	$9.45\ t\ km^{-2}\ a^{-1}$	Largely dissolved SiO_2
Sogndal, Norway	$13\ meq\ m^{-2}\ a^{-1}$	
Lake Gardsjon, Sweden	$40–80\ meq\ m^{-2}\ a^{-1}$	
Germany	$20–230\ meq\ m^{-2}\ a^{-1}$	Forest soils
Scotland	$19.49\ meq\ m^{-2}\ a^{-1}$	

Source: Compiled from information in: Colman, S.M. and Dethier, D.P. (eds) 1986: *Rates of chemical weathering of rocks and minerals*. London: Academic Press; Bain, D.C., Mellor, A., Wilson, M.J. and Duthie, D.M.L. 1990: Mineral weathering studies in Scandinavia. In Mason, B.J. (ed.) *Surface waters acidification programme*. Cambridge: Cambridge University Press.

in measurements of ion loss. In reality, however, a catchment may contain a young forest that is accumulating cations such as calcium, or old woodlands that are net losers. The assumption of a steady state may be a source of error.

- *It is unrealistic to convert chemical losses into ground surface lowering*. This is because chemical weathering brings about an increase in volume that is hard to measure and, in humid regions, surface lowering is in any event brought about by creep and other mass movements.
- *The source of measured water may affect the significance of the results*. Water that has been in contact with the regolith for a relatively short time will have a low chemical load; long contact times will on the other hand yield high loads. Some investigations in Scandinavia have shown that catchment studies may give lower values than soil profile studies. This may be because most runoff water in the areas examined follows shallow, rapid pathways.

In spite of these problems, many studies have been made of weathering rates using the geochemical approach, and some of the results are given in Table 9.12.

Weathering changes in individual rocks and minerals

The weathering of a rock fragment or individual mineral may bring about a measurable alteration over time. In the case of a rock the alteration may consist of the development of a weathered rind whose thickness can be measured. A more indirect approach has been to measure the velocity of compression waves generated by a hammer blow against, for example, granite boulders. The velocity reduces with time due to the development of microfractures resulting from weathering. In both cases, assuming a numerical age for the deposition of the fragment is known, a weathering rate can be calculated. For the western USA, studies have shown that weathering rinds have developed on basaltic and andesitic clasts at an average rate of $5\,\mu m\,ka^{-1}$ over the last 500 000 years.

The central problem with this approach is the relationship between the degree of weathering and time. This has two complications.

- The rate of weathering tends to decrease with time, and so the relationship is exponential. For weathered rinds it has the form

$$d = kt^{\frac{1}{n}} \tag{9.1}$$

  where d is rind thickness; t is time; k is a constant related to age, where known; n is a number ≥ 1 (<2 for rinds forming by a combination of hydrolysis, oxidation and solution).
- The degree of weathering is a function of factors such as temperature and water availability as well as of time. These factors may be difficult to disentangle.

The same approach has been used at the scale of the individual mineral, e.g. the alteration of hornblende. This mineral weathers relatively easily, and its cleaved structure encourages the development, by 'etching', of ragged ('cockscomb') grain edges. The degree of etching can be measured. (*See* Fig. 9.9.)

Where the age of the containing sediment is known, a logarithmic model relating the degree of etching to the depth of the particle may be applied. It is

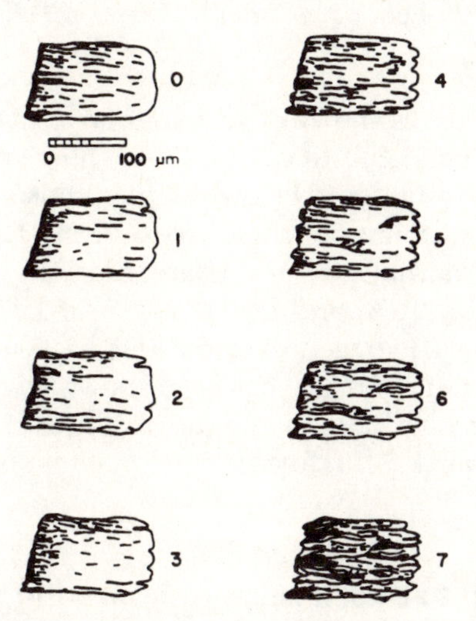

FIGURE 9.9 The progressive etching of hornblende grains through etching classes. (*Source*: Locke, W.W. 1979: Etching of hornblende grains in arctic soils: An indicator of relative age and palaeoclimate. *Quaternary Research* **11**, 197–212.)

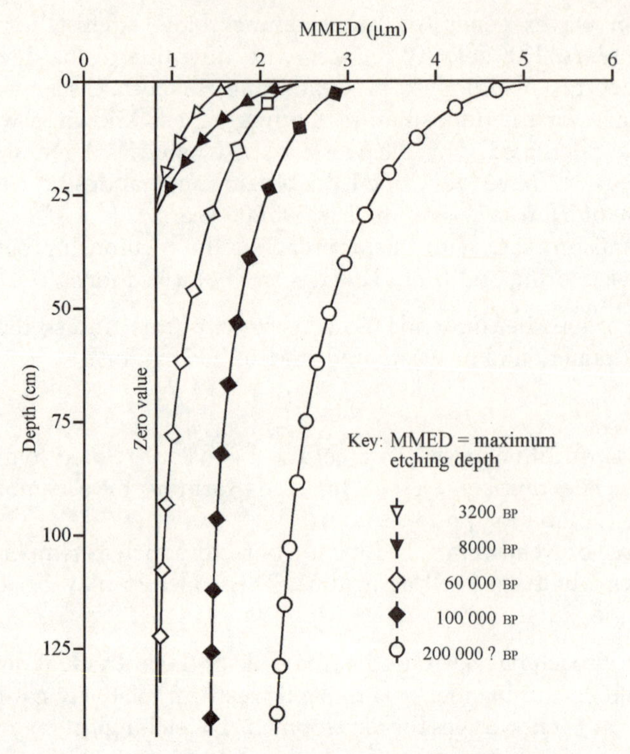

FIGURE 9.10 Summary of some etching results for Arctic soils. (*Source*: Locke, W.W. 1986: Rates of hornblende etching in soils on glacial deposits, Baffin Island, Canada. In Colman, S.M. and Dethier, D.P. (eds) *Rates of chemical weathering of rocks and minerals*. London: Academic Press, 129–45.)

then possible to plot curves showing the variation of the degree of weathering with depth (*see* Fig. 9.10). In summary, etching rates are logarithmic functions of both time and depth below the regolith surface.

Studies show that maximum etching rates during the Holocene (i.e. the last 10 000 years) for hornblende grains are generally less than 1 $\mu m\,ka^{-1}$ in southwest Montana, USA, and less than 0.5 $\mu m\,ka^{-1}$ for Baffin Island, northern Canada. Rates averaged over the last 100 000 years are less than 0.1 $\mu m\,ka^{-1}$. The effect of grain size has been shown by studies in the northern Rockies, where etching rates of 0.02 $\mu m\,ka^{-1}$ in coarse-textured granitic materials and of 0.21 $\mu m\,ka^{-1}$ in fine-grained material have been calculated for a 12 000 year period.

9.6.3 Relationship between rates measured in the field and in the laboratory

Most studies have shown that chemical weathering rates in the field, as measured by the mass balance method (based on a relationship between

TABLE 9.13 Dissolution rates for various minerals under laboratory and field conditions. The field sites were experimental plots on podzolic soils in E. Maine, USA.

Mineral	Dissolution rate (at pH 4.5) (pmol $m^{-2} s^{-1}$)		Ratio laboratory/field
	Laboratory	Field	
Plagioclase	13.6	0.052	262
Potassium feldspar	16.7	0.052	321
Muscovite	6.3	0.021	300
Biotite	3.8	0.009	422
Chlorite	3.3	0.010	330
Hornblende	1.4	0.003	467

Source: Extracted from table in Swoboda-Colberg, N.G. and Drever, J. 1993: Mineral dissolution rates in plot-scale field and laboratory experiments. *Chemical Geology* **105**, 51–69.

input and output), are one to three orders of magnitude slower than those determined by laboratory investigations (*see* Table 9.13). There are several possible explanations for this discrepancy.

- *Temperature differences between laboratory and field.* It is well known that weathering rate is influenced by temperature. Studies have, however, shown that the maximum effect of likely temperature variation (perhaps up to 30°C) on the rate is probably no greater than a factor of five. This can only account for a small part of the overall discrepancy of between one and three orders of magnitude.
- *The insulation of minerals grains by surface coatings in field situations.* This is unconvincing as observations suggest that naturally weathered silicate mineral grains are extensively etched, and so offer complete access to percolating fluids.
- *Reduced contact between water and minerals in field situations.* A 'wetting' event, such as a period of rain or snow melt, typically brings only a fraction of the available mineral surface into contact with water. In addition the movement of water through the vadose zone (i.e. above the water table) is both spatially very restricted and limited in amount. Experiments have shown that a wetting event may saturate as little as 0.2–15 per cent of the total cross-section through which water may flow. Such values may be achieved at depths of only several tens of centimetres in finely textured soils. Further, theoretical models have suggested that only about 10 per cent of the actual soil moisture takes part in flow.

It appears then that hydraulic factors are the most important in explaining the discrepancy between field and laboratory rates of chemical weathering.

Readings for further study

General reviews

Brewer, R. 1964: Mineral stability and weathering. In *Fabric and mineral analysis of soils*. New York: John Wiley, Chapter 5. This classic reference is still helpful as a survey of some of the earlier ideas about mineral stability and the measurement of weathering intensity.

Colman, S.M. and Dethier, D.P. 1986: An overview of rates of chemical weathering. In Colman, S.M. and Dethier, D.P. (eds) *Rates of chemical weathering of rocks and minerals*. London: Academic Press, 1–18. Summarises important modern ideas. These include the influence of weathering fluids and the nature of reactions on weathering rates, and the measurement of rates by mineral, rock and water-chemistry studies.

More specialised studies

You will find much valuable material in the various chapters of the book edited by Colman and Dethier (above), but take care not to become submerged by detail. You will also find that the various papers referenced within this chapter will expand the ideas we have outlined. Among the more accessible articles are the following.

April, R., Newton, R. and Coles, L.T. 1986: Chemical weathering in two Adirondack watersheds: past and present-day rates. *Geological Society of America Bulletin* **97**, 1232–8. Shows how the loss of minerals and elements from soils, and studies of surface-water chemistry, can lead to estimates of weathering rates.

Cooke, R.U., Inkpen, R.J. and Wiggs, G.F.S. 1995: Using gravestones to assess changing rates of weathering in the United Kingdom. *Earth Surface Processes and Landforms* **20**, 531–46. A calculation of weathering rates from the height differences between lead lettering and adjacent marble on tombstones at sites in Swansea, Portsmouth and Wolverhampton.

Swoboda-Colberg, N.G. and Drever, J. 1993: Mineral dissolution rates in plot-scale field and laboratory experiments. *Chemical Geology* **105**, 51–69. Shows experimentally that dissolution rates in the laboratory are higher by a factor of 200–400 than those in the field, where there is poor contact between minerals and weathering solutions, pH near mineral surfaces is relatively high, and high aluminium concentrations reduce weathering rates.

Wasklewicz, T.A. 1994: Importance of environment on the order of mineral weathering in oliving basalts, Hawaii. *Earth Surface Processes and Landforms* **19**, 715–34. A discussion of evidence that Goldich's classic sequence of mineral susceptiblity to weathering may vary with the nature of the biochemical environment.

10

PRODUCTS OF WEATHERING

In the last chapter you studied the ideas of weathering intensity and of the rate at which alteration may proceed. A consequence of rate and intensity is the weathering product. You have already seen how this may vary in a geographic sense, i.e. over global distances, and the ideas are developed further below (*see* Box 10.3). A focus in this chapter is on the vertical character of the weathered product, with special reference to the development of distinct weathering sequences. Locally, weathered products of high economic value may emerge.

The aim then in this chapter is to examine

- the nature of the weathered product;
- its spatial arrangement, with special reference to vertical organisation, as expressed in *laterites, silcretes* and *calcretes;*
- some economic consequences of weathering.

10.1 Nature of the weathered product: the results of chemical weathering

The various processes of chemical weathering decompose the relatively stable primary minerals to yield a large array of alteration, or secondary, products. These are sometimes called *neoformations* ('neo' means 'new') to emphasise their changed character. Some of these are soluble, and so are quickly lost from the immediate weathering environment; others are quite readily precipitated, such as calcite in limestone voids, while yet others are relatively insoluble. The focus in this chapter is on the two latter categories, which include:

- the crystalline phyllosilicate minerals ('phyllo' means 'leaf-like');
- poorly ordered aluminosilicates;
- oxides and hydroxides of iron, aluminium and manganese.

10.1.1 Crystalline phyllosilicate clays

These are the clay minerals as narrowly defined, and which make up the largest group of alteration products. They are characterised by small size (<2 μm diameter) and crystalline nature, expressed by a generally platy structure (hence 'phyllosilicate'). The basic principles are discussed in Chapter 2. You can understand their main features if you focus on two themes:

- the nature of the three-dimensional sheets that constitute the minerals;
- the bonding between sheets, and between collections of sheets (called layers).

Structure of the sheets

There are two basic types of sheet that form the fundamental building blocks of clay minerals: one made from linked SiO_4 tetrahedra, the other from linked octahedra.

The tetrahedral sheet
This consists (*see* Fig. 10.1) of an extended array of tetrahedra. Each tetrahedron is linked with its neighbours by sharing its basal oxygens, so forming a hexagonal (six-sided) mesh pattern (*see* Fig. 10.2). The two-dimensional array of basal oxygens is an example of a *plane*. All the fourth oxygens of the tetrahedra (apical oxygens) point in the same direction, at right angles to the sheet. The enclosed ion of each tetrahedron is normally Si^{4+}, but there may be some replacement by Al^{3+} or Fe^{3+}. This clearly has implications for the overall charge.

The octahedral sheet
This array is made up of individual octahedra, each consisting of six oxygen or hydroxyl ions, and linked laterally by sharing the octahedral edges (*see* Fig. 10.3).

Each octahedron has a net negative charge of 2. To maintain neutrality within the sheet, only two out of three octahedra need to be occupied by a trivalent cation such as Al^{3+}, giving a *dioctahedral* sheet. This gives the *gibbsite* structure which is $Al(OH)_3$. Alternatively, and still to maintain neutrality, all

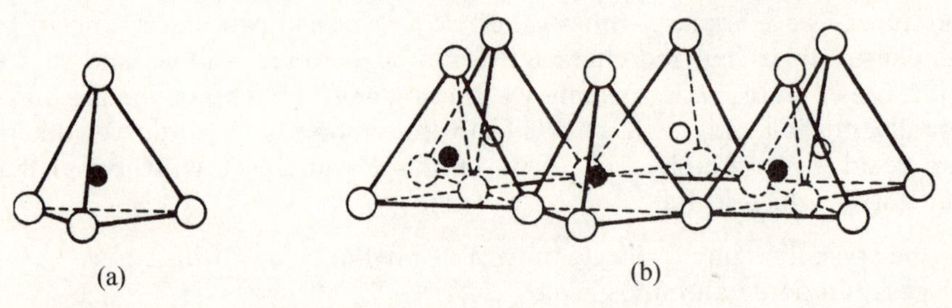

FIGURE 10.1 (a) A single SiO_4 tetrahedron and (b) the sheet structure of SiO_4 tetrahedra. ○ and ◌ = oxygens; ○ and ● = silicons.

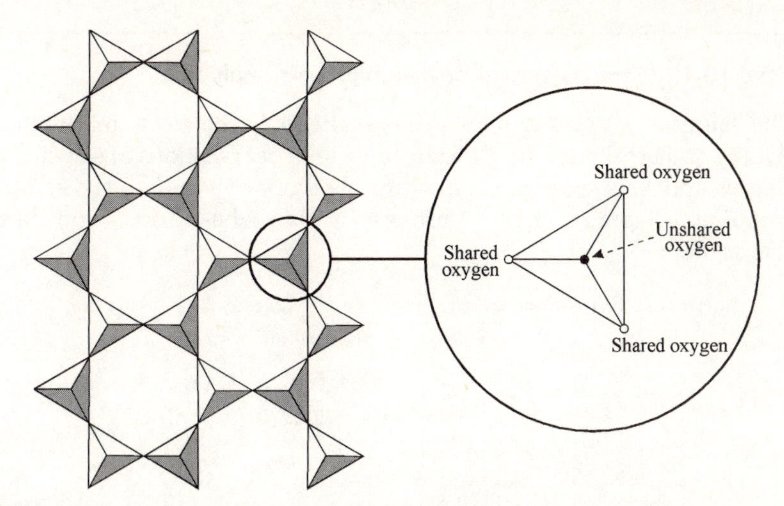

FIGURE 10.2 The linking of three tetrahedra to produce a phyllosilicate sheet structure. The detail identifies shared and unshared oxygens.

the octahedra may be occupied by a divalent cation, e.g. Mg^{2+}, thus giving a *trioctahedral* sheet. In this case the result is *brucite,* $Mg(OH)_2$.

Linkage between sheets and layers, and some important resulting minerals (*see* Box 10.1)

So far we have seen how a number of planes of atoms (e.g. the basal oxygens of linked tetrahedra) are linked to form sheets (e.g. the octahedral sheet). Now we must examine how sheets are joined to form *layers,* and how these are combined to form *unit structures.*

There are two main ways in which tetrahedral and octahedral sheets are linked to form layers (*see* Fig. 10.4).

Linkage by combining to form a '1:1 layer'

In this case a single tetrahedral sheet and a single octahedral sheet are linked, hence the description '1:1 layer'. The paired structure is about 7 Å thick. The

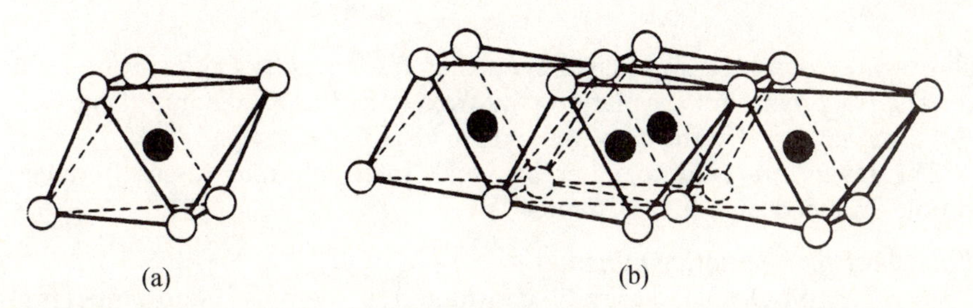

FIGURE 10.3 (a) A single octahedron and (b) the sheet structure of linked octahedra. ○ and ◌ = hydroxyls; ● aluminiums, magnesiums, etc.

Box 10.1 Structures of some clay minerals

The 'simple' diagrams show the relationship between tetrahedral ( ) and octahedral ( ) sheets. The usual positions of structural charge and of exchange cations are shown by + and – signs. The 'detailed' diagrams give examples of how the various ions are arranged.

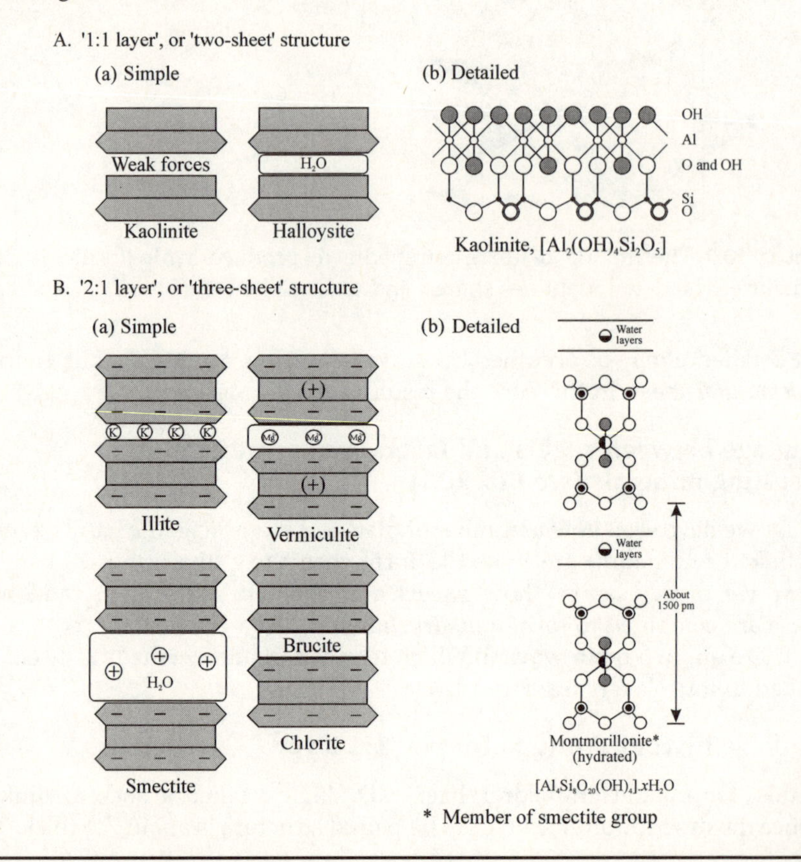

sheets are combined in such a way that the oxygens at the tips of the tetrahedra project into a plane of hydroxyls in the octahedral sheet. The oxygens replace two-thirds of the hydroxyls.

The layers are held together by weak intermolecular forces (induced dipole–induced dipole; *see* Chapter 1).

Important representative minerals

Kaolinite ($Al_2Si_2O_5(OH)_4$) may be considered as the most important representative of the 1:1 type. In this case only two-thirds of the possible positions for aluminium in the octahedral sheet are occupied. The overall charge of the

Two-sheet or 1:1 layer, clay minerals

Layer { Silicate sheet (tetrahedral sheet)
$Al(OH)_3$ or $Mg(OH)_2$ sheet (octahedral sheet)

Three-sheet or 2:1 layer, clay minerals

Layer { Silicate sheet (tetrahedral sheet)
$Al(OH)_3$ or $Mg(OH)_2$ sheet (octahedral sheet)
Silicate sheet (tetrahedral sheet)

FIGURE 10.4 How sheets are arranged to form '1:1' and '2:1' layers.

various planes in the structure is neutral.

SiO_4 tetrahedra	$6O^{2-}$	12−
	$4Si^{4+}$	16+
Common plane	$4O^{2-} + 2(OH)^-$	10−
Octahedra	$4Al^{3+}$	12+
	$6(OH)^-$	6−
		0

Kaolinite is built of successive layers, superimposed in such a way that the oxygens at the base of one layer are paired by close approach to hydroxyl ions at the top of its neighbour (O–(OH) bonds). This requires some distortion. The forces holding the layers together are of the 'weak intermolecular' type. Water does not enter between layers, and so kaolinite is 'non-swelling'.

The variety *halloysite* ($Al_2Si_2O_5(OH)_4 \cdot 2H_2O$) has a single sheet of water molecules between its structural sheets, forming hydrogen bonds with the sheets of oxygens or hydroxyls on either side. Stacking is consequently disordered, and the interlayer distance increases from about 7.2 to 10 Å.

Serpentine ($Mg_3Si_2O_5(OH)_4$) is similar to kaolinite, except that a sheet of brucite ($Mg(OH)_2$) rather than a sheet of gibbsite ($Al(OH)_3$) is present. It is non-swelling.

Linkage by combining to form a '2:1 layer'

The basic structure consists of three sheets, an octahedral sheet sandwiched between two silica tetrahedral sheets, to form a layer about 10 Å thick. The apical oxygens point towards the central octahedral sheet and substitute for two-thirds of the hydroxyls of the octahedra. The layers are held together by

- cations, as in illite and vermiculite;
- a *mixed brucite sheet*, $Mg(OH)_2$, with some Mg^{2+} replaced by Al^{3+}, as in chlorite.

Important representative minerals

It is convenient to recognise four broad subdivisions of the 2:1 layer group as far as weathering is concerned. The subdivisions differ typically in terms of the degree of ion substitution in the tetrahedra, and in the types of interlayer ion.

Illite

This clay mineral differs from a 'standard' 2:1 layer structure (e.g. mica, a primary mineral) in the following respects.

- There is only a modest (about one-sixth) substitution of Al^{3+} for Si^{4+} in the tetrahedra. As a consequence, the silicon to aluminium ratio is relatively high, and the net charge deficiency (loss of positive charge) is reduced to about 1.3 per unit cell.
- The interlayer cations, which bond the layers together, consist dominantly of K^+, plus Ca^{2+}, Mg^{2+} and H^+. They are relatively few in number, because of the modest charge deficiency, and so forces between the layers are weaker.

The presence of interlayer potassium ions in illite prevents the entry of water and other cations, and so illite has a low cation exchange capacity (10–40 meq ($100\ g^{-1}$)). This is less than those for halloysite, montmorillonite and vermiculite, but actually greater than for most kaolinites.

Smectite group

For this group a central sheet of gibbsite ($Al(OH)_3$) has been modified by the replacement of two of every three (OH) ions by apical oxygens. Substitutions by ions of lower valency occur in both octahedral sites (e.g. of Al^{3+} by Fe^{2+}, Mg^{2+} and Zn^{2+}) and tetrahedral sites (e.g. of Si^{4+} in SiO_4 by Al^{3+}). The electronic imbalance is corrected by the presence of a small number of interlayer cations, typically Na^+ or Ca^{2+}. These are mostly exchangeable. In the case of *montmorillonite* the replacement of Al^{3+} by Mg^{2+} in the octahedral sheet is typical, as is the occurrence of Na^+ as the interlayer cation.

Water is readily absorbed between the structural layers, hence the description 'swelling clays' for *smectites* (including montmorillonite). The water may be organised into layers, whose number is affected by the types of interlayer cation. Calcium montmorillonites (the most common) have two layers with a variable sodium ion content. A consequence is that basal spacings vary between 10 and about 21 Å.

Vermiculite

The central sheet of vermiculite should be regarded as a brucite with octahedrally coordinated Mg^{2+} and Fe^{2+} ions. Appreciable substitution by Al^{3+} and Fe^{3+} occurs. The brucite sheet lies between two sheets of linked tetrahedra, whose apical oxygens point inwards. These apical oxygens replace two of every three hydroxyls on each side. There is some replacement of tetrahedral silicon by Al^{2+} or Fe^{3+}, which is compensated by the introduction of interlayer cations, mainly Mg^{2+}. Water molecules are also formed between the layers.

In a naturally occurring magnesium vermiculite a pair of water sheets is found, bonded by an exchangeable cation that lies between them ($H_2O–Mg^{2+}–H_2O$). Each molecule of a water sheet is linked by a hydrogen bond to an oxygen on the silicate layer surface. The presence of water means that vermiculite is a swelling clay. The basal spacing is about

TABLE 10.1 Layered silicates

Main group	Subgroup	Examples
Two layers (1 : 1)		
Kaolinite	Kaolinite	Kaolinite, halloysite
Kaolinite	Serpentine	Chrysotile, greenalite
Three layers (2 : 1)		
Pyrophyllite	Pyrophyllite, talc	Pyrophyllite, talc
Smectite	Dioctahedral	Montmorillonite, beidelite
Smectite	Trioctahedral	Saponite, sauconite
Vermiculite	Dioctahedral	Vermiculite
	Trioctahedral	Vermiculite
Mica	Dioctahedral	Illite, muscovite, sericite
Mica	Trioctahedral	Biotite
Chlorite	Trioctahedral	Corundophite, pseudothuringite, talc-chlorite

14.4 Å, falling with dehydration (loss of water) to about 9 Å. The layer occupied by water molecules and hydrated cations is normally about 5 Å thick under atmospheric conditions. The cation exchange capacity (100–260 meq $(100\ g^{-1})$) is greater than that of smectites, and is the highest of all clay minerals.

Chlorite minerals

These are also 2:1 layer minerals, but differ from those discussed above in that the interlayer space between the 2:1 structures (which constitute 'mica-like' layers) is occupied by a further octahedral sheet, a brucite-like layer. This consists of hydroxyls coordinated to magnesium and aluminium. The mica-like layer (an octahedral sheet sandwiched between two tetrahedral sheets) is unbalanced by the substitution of Al^{3+} for Si^{4+}, and this deficiency of charge is balanced by an excess charge in the brucite-like sheet as a consequence of the substitution of Al^{3+} for Mg^{2+}.

The bonding between the layers is partly electrostatic in character as a consequence of substitution within the lattice. The strongly bound brucite sheet ensures that no hydration, and so swelling, of this mineral occurs.

Table 10.1 provides a useful list of layered silicates, some of which are primary minerals (e.g. muscovite) and others (e.g. illite) are secondary.

10.1.2 Non-crystalline and poorly ordered aluminosilicates

These minerals lack the well-defined unit cell of the phyllosilicates and so are difficult to identify. Three varieties are important in weathering environments.

- *Amorphous silica* (SiO_2) is quite common in the upper part of the regolith in cool temperate environments. Silica is taken up by plant roots and then precipitated around cell walls. Finally, on the death of the plant, it is returned to the regolith. The term 'amorphous' means 'lacking in crystalline structure'.

- *Allophanes* are members of a series of hydrous aluminosilicates, which contain varying amounts of SiO_2, Al_2O_3 and H_2O. The ratio of silicon to aluminium ranges from 1 to 2. The composition of a typical allophane is $2SiO_2 \cdot Al_2O_3 \cdot 3H_2O$.
- *Imogolite* is a poorly-crystalline fibrous mineral, having the formula $SiO_2 \cdot Al_2O_3 \cdot H_2O$. Like allophane, it was originally identified in weathered volcanic glasses in tropical and subtropical environments, but is now thought to occur in small quantities more widely.

10.1.3 Metallic oxides, oxohydroxides and hydroxides

Metallic oxides, oxohydroxides and hydroxides, especially those of iron, aluminium, magnesium and manganese, are important constituents of some regoliths. They are residual compounds resulting from weathering. *Oxides* are compounds of metals and oxide ions (O^{2-}), *hydroxides* are compounds of metals and hydroxide ions (OH^-) and *oxohydroxides* are compounds which contain a mixture of oxide ions (hence 'oxo') and hydroxide ions (hence 'hydroxide'). With strongly basic metals, e.g. magnesium, only hydroxides will tend to form as oxides are rapidly hydrated. For less basic metals, e.g. iron and aluminium, all three types of compound are found in nature, and may form massive deposits. Some important examples of the various compounds are given in Table 10.2. You will find it convenient to study these important secondary minerals by considering each of the main metals and its compounds.

Iron

Iron forms **oxides,** two of which occur as residual compounds. *Hematite* is the most important. It has the ideal composition Fe_2O_3 and therefore contains mainly ferric ions (Fe^{3+}) and oxide ions (O^{2-}). In addition, normally small

Table 10.2 Some compounds formed from important metals

Metal	Oxide	Oxohydroxide	Hydroxide
Iron, Fe	Hematite, α-Fe_2O_3 Maghemite, γ-Fe_2O_3	Goethite, β-FeO(OH) Lepidocrocite, γ-FeO(OH) Ferrihydrite (see text) Limonite (see text)	
Aluminium, Al	Corundum, α-Al_2O_3	Diaspore, α-AlO(OH) Boehmite, γ-AlO(OH)	Gibbsite, γ-$Al(OH)_3$
Magnesium, Mg			Brucite, $Mg(OH)_2$
Manganese, Mn	Pyrolusite, MnO_2		

Note: the prefixes α (alpha), β (beta) and γ (gamma) denote crystalline modifications (see text).

amounts of manganese (Mn^{2+}) and ferrous (Fe^{2+}) ions are also present. The structure of hematite (Fe_2O_3) has been described in Chapter 2. It is widely found in tropical environments, and may result from the weathering by dehydration of *goethite* ($FeO(OH)$):

$$2FeO(OH) \longrightarrow Fe_2O_3 + H_2O$$

The blood-red colour of hematite makes it a striking component of many highly leached tropical regoliths, especially those developed on basic rocks which are poor in the acidic oxide SiO_2 but may be relatively rich in calcium.

Maghemite is another form of ferric oxide, where the crystalline modification is called γ-Fe_2O_3. Hematite, α-Fe_2O_3, has only octahedrally coordinated Fe^{3+} ions, while in γ-Fe_2O_3 one-third of the Fe^{3+} ions are found in tetrahedral sites. The labels α- and γ- are used to distinguish the two structures (*see* Chapter 2). The presence of maghemite in some soils is perhaps due to the heat of burning vegetation, which may cause dehydration of oxides and oxohydroxides as well as crystalline modification of the Fe_2O_3 to the maghemite form.

Iron may also form **oxohydroxides** (sometimes called oxyhydroxides), a class which includes several important compounds. *Goethite* is a yellowish-brown to red hydrated oxohydroxide. It has a structure similar to that of boehmite (see below), except that Fe^{3+} replaces Al^{3+} in the octahedral positions. It is the most abundant iron-containing oxohydroxide in the regolith, especially in temperate regions. It is normally formed by the hydrolysis of Fe^{3+} ions released by the oxidation of iron-containing minerals such as pyrites (FeS_2). When dehydrated it yields hematite (see above).

Lepidocrocite is a brownish to red oxohydroxide which is structurally similar to boehmite, but with iron-centred oxygen octahedra linked in chains. It may form through the oxidation of Fe^{2+} ions, especially in poorly drained regoliths. On dehydration it gives maghemite (see above).

Ferrihydrite is a poorly ordered oxohydroxide. Its high content of water of hydration gives it a composition between $Fe_4(OH)_{12}$ and $Fe_5O_3(OH)_9$. It may be the major iron-containing secondary mineral in many iron concentrations in poorly drained soils, and in the iron-rich horizons of podzols (*see* Box 10.2). It is, however, a metastable mineral, and therefore only slowly transforms to hematite in the tropics and to goethite in temperate regions.

Limonite consists of poorly crystalline goethite or lepidocrocite and absorbed water, i.e. $FeO \cdot OH \cdot nH_2O$. The name is best restricted to field use when it refers to a widely found alteration product of iron-bearing minerals, especially in wet environments under temperate conditions. Its colours, ranging from yellow to brown, orange-brown and brownish black, contrast with those of hematite.

Aluminium

The only **oxide** of aluminium is the hard and chemically resistant mineral *corundum*. It is rare. There are two forms of aluminium **oxohydroxide**. *Diaspore* ($AlO(OH)$) is the alpha form. It crystallises from an aluminium silica gel produced by the hydrolysis of silicates under humid tropical conditions, and is mainly found in bauxites. *Boehmite* (AlO(OH)), the gamma form, occurs as the main constituent of some bauxite deposits, and may be found with diaspore.

Manganese

The most common **oxide** of this metal is *pyrolusite* (MnO_2) which, together with *hausmannite* (Mn_3O_4), is formed by the weathering of manganese-containing silicates. Commercial deposits occur quite widely on land (e.g. former USSR, South Africa, Australia, China and Brazil), but of particular interest is the accumulation on the sea floor of colloidal particles of the oxides, which aggregate to form manganese nodules.

Magnesium

The **hydroxide** *brucite* ($Mg(OH)_2$) is the most important magnesium compound in weathering environments. This basically white mineral is often discolored brown or green due to the presence of iron (Fe^{2+}) ions or manganese (Mn^{2+}) ions.

10.2 Nature of the weathered product: results of mechanical weathering

Remember that mechanical weathering processes do not involve any chemical alteration of the weathering material. The products are therefore chemically unchanged, and consist essentially of broken fragments of the original rock. However, matters are a little more complicated than this. The character of the weathered product is strongly influenced by the properties of the parent rock and, in the case of frost action, by the nature of the weathering process. In addition, the product may contain chemically altered material, derived either from the parent rock itself or from the alteration of mechanically derived debris. These ideas are developed below.

10.2.1 Influence of parent-rock properties

You will find it helpful to distinguish between two types of mechanically weathered regolith, each of which is related to the physical properties of the parent rock.

Granular regolith

Here the regolith is dominated by granules (particles between 2.0 and 4.0 mm diameter), together with sands (0.0625–2.0 mm) and silts (0.008–0.0625 mm). Generally the size reflects the textural character of the parent rock.

For strongly *crystalline* rocks, breakdown may take place along mineral boundaries, and so the resulting coarse grains are monomineralic ('one mineral'). However, a rather more common situation involves failure along other crack systems, and so the released particles are polymineralic ('many minerals'). The term '*grus*' is used to describe a coarse granular regolith derived from granitic rocks.

Sedimentary rocks tend to yield generally smaller particles and so a finer regolith. Experimental work (especially in France) has provided evidence about the relationship between the type of limestone and the size of particle released by frost weathering. Chalk tends to yield a high percentage of fine material (<0.05 mm diameter). In contrast a dense lithographic limestone (a compact limestone used in the printing process called lithography) releases larger particles, between 1 and 2 cm, with little fine material.

Rather larger debris may result from the breakdown of *metamorphic* rocks. For example, the quartz-mica schist (a platy metamorphic rock) of Signy Island (South Orkney group, Antarctica) yields a dominant coarser component that is about 70 per cent larger than granular size.

Debris consisting of small flakes, perhaps 1–10 mm thick, may accumulate where rocks are *massive*, i.e. weakness planes are relatively far apart. This is particularly the case for granites and has also been described for volcanic rocks, for example near Lake Magadi, Kenya, where salt crystallisation has led to the flaking of lava surfaces.

Regolith consisting of coarse, angular fragments

This second type of regolith is produced when larger-scale planes of weakness are exploited by mechanical weathering processes. The result is a coarse deposit whose component fragments may be up to several metres in length and which has a high proportion of voids. When such deposits are formed by frost action, they constitute blockfields and screes (*see* Chapter 11).

10.2.2 Contribution of environmental conditions

For frost weathering, fragment size may be affected by the variation of seasonal freeze–thaw temperatures. Relatively small fragments may be produced in regions where the depth to which freeze–thaw temperatures descend is shallow, for example in the Sør Rondane Mountains of Antarctica. Larger fragments may be produced in regions experiencing deep freeze–thaw penetration, such as northern Scandinavia and Baffin Island.

10.2.3 Contribution of chemical alteration processes

The nature and thickness of a regolith produced by mechanical breakdown may be affected by the degree to which the parent rock had been chemically altered, and by the amount of chemical weathering that has affected the regolith itself. In both cases the regolith may have a 'chemical' component.

Examples of the first influence have emerged from both field and laboratory studies. Intense granular disintegration has taken place over a distance of about 7 km on a medium- to coarse-grained granite near Göteborg, Sweden. Breakdown has been accelerated because of the presence of a swelling clay mineral (montmorillonite), itself due to the action of hot fluids during an earlier geological period, along the grain boundaries. Laboratory studies have provided evidence that frost action is more effective when preliminary alteration has occurred. This has been observed for slightly altered granites and schists, which yield larger volumes of finer material than unaltered rock.

A mechanically weathered regolith may be affected by later chemical processes. This is particularly the case when mechanical weathering largely ceased some time ago. An example is provided by the chemical alteration during the Holocene (the last 10 000 years) of regoliths produced by frost action during, at least, the previous glacial episode. Such regoliths show a number of chemically produced features.

- *Colour changes* at the surface of rock fragments, which are of two varieties. Rusty colours are common on basic rocks, and are often due to the chloritisation (i.e. conversion to the mineral chlorite, a layered silicate) of micas. This involves an enrichment in iron (which gives the rusty colour) and a depletion in silicon, aluminium, magnesium and potassium. Elsewhere a greyish-white colour is found which is largely due to the presence of iron in the reduced state at the rock surface.
- *Coloured crusts* may occur where rock surfaces are wetted and dried. Such crusts are about 0.01–0.05 mm thick, and consist of amorphous silicon- and iron-rich compounds. The movement of iron and other metal ions towards the surface occurs during drying episodes.

10.3 Spatial variation of chemically weathered regoliths

Observations show that there is often great variation in the nature of a chemically weathered regolith. This variation may be expressed both vertically, at a locality, and horizontally over the Earth's surface. You will study these ideas further in the following sections.

Altered regoliths typically show vertical variation simply because the intensity of weathering tends to change with depth. Such variation is often best developed in humid tropical environments. In addition, the nature of the regolith tends to vary with broad climatic zone over the Earth's surface.

10.3.1 Vertical variation

As the intensity of weathering tends to decrease with depth, a layered arrangement typically develops. The layers are called *weathering zones* or *horizons*. If two or more zones occur, a vertical section through them makes up a *weathering profile* (*see* Fig. 10.5).

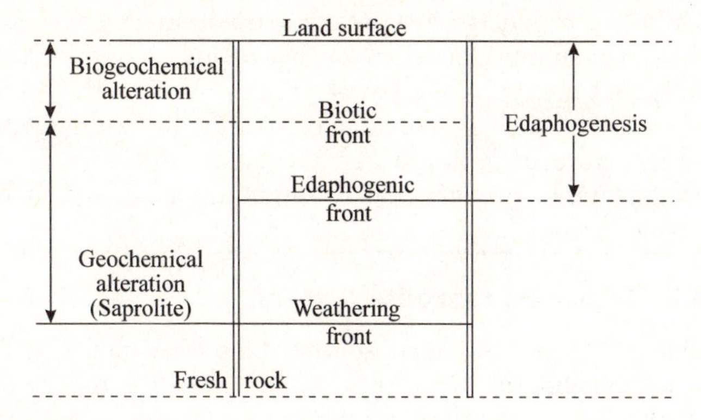

FIGURE 10.5 Some zones and junctions in an idealised weathering profile for humid regions (see text for explanation).

Ideally, weathering begins when fresh rock is exposed at a land surface (*see* Fig. 10.5). The junction between the developing regolith and fresh rock is the *weathering front*, which advances downwards as weathering proceeds. Immediately above it is a zone consisting of *saprolite*, which is altered material possessing the structure of the original rock.

At some distance behind the weathering front, alteration becomes sufficiently advanced for the rock structure to be destroyed. The front at which the transition occurs is the *edaphogenic* (soil-producing) *front*, above which soil-forming processes dominate (*see* Box 10.2). At a certain distance above this front, i.e. at the *biotic front*, biological activity starts to dominate.

A useful simplification is to divide the regolith into a lower zone of saprolite, characterised by *geochemical alteration* (i.e. the action of non-biological solutions), and an upper zone, the biological soil, where *biogeochemical alteration* occurs.

10.3.2 Horizontal variation

Generally, regoliths developed by the chemical weathering of aluminosilicate rocks under cool, humid conditions tend to be silica rich, since alumina is quite readily removed. Under warmer, humid conditions silica is more easily lost and regoliths are alumina enriched to varying degrees. These ideas are developed further in Box 10.3. The behaviour of iron is broadly similar to that of aluminium.

Locally, and under suitable conditions, calcium or silicon enrichment may occur, giving *calcrete* and *silcrete* respectively.

10.4 THE LATERITIC PROFILE

Regoliths tend to be thickest (perhaps 200 m) and most complicated in humid intertropical areas. The term *laterite* may be used to describe all kinds of tropical weathering products, i.e. those made up of mineral assemblages that may include iron or aluminium oxides, oxohydroxides or hydroxides, together with kaolinite and quartz. A variety of profiles occurs, whose character is determined by factors such as present and past climate, parent rocks, surface slope and the nature of groundwater circulation. You should study first some simplified lateritic profiles, before examining the more detailed types.

Box 10.2 Regolith, saprolite and soil

The *regolith* is the mantle of unconsolidated material, including the soil, which rests on solid, unaltered rock. It may consist of the products of physical and/or chemical weathering, or of loose sediments laid down by agents such as wind and water.

Saprolite is a term for rock that has been strongly altered by chemical weathering. It lies below the soil, and is found especially in humid tropical regions. *Soil* refers to the upper part, perhaps 0.3 to 2.0 m or more thick, of the regolith. It is identified by the presence of organic activity. The topmost layer is the *organic* or *O horizon*, dominated by plant matter in various states of decay. *Mineral horizons* develop below, mainly as a result of two major processes. *Eluviation* is the removal of clays and solutes from a horizon, and *illuviation* is their accumulation, usually in a lower horizon. There are four main mineral horizons, each identified by a capital letter (*see* Fig. 10.6), and distinguished by properties such as colour and degree of weathering. Lower-case letters may be used to subdivide horizons, e.g. Ea denotes a bleached layer in a podzol (see below). The vertical sequence of horizons forms the *soil profile*. The nature of any individual soil is due to the interaction of five factors: climate, organisms (especially vegetation), topography, parent material and time. When the first two factors operate for a long time they give rise to *zonal soils*, which roughly coincide with the Earth's major climatic zones. Three types of zonal soil reflect contrasting weathering environments.

- *Podzols* develop under cool, humid, acid conditions, where water percolates readily. They show strong eluviation, resulting in pale, ashy E horizons, and marked illuviation in the B horizon where iron typically accumulates.
- *Oxisols* are thick soils which result from long-continued weathering under humid tropical conditions. Their red, yellow and yellow-brown colour is due to the presence of much iron in its two oxidation states. The organic matter is low due to the high rate at which it is broken down.
- *Chernozems* occur in semi-arid mid-latitude regions. They are rich in soluble salts, notably of calcium (Ca^{2+}), as there is not enough rainfall for significant eluviation to occur.

Box 10.2 *Continued*

Horizon	Description
O	Organic matter in various stages of decomposition from fresh leaf litter to humus
A	A dark-coloured mixture of organic and mineral matter
E	Horizon of maximum eluviation of clays and solutes; light coloured
B	Illuvial horizon, where oxides of aluminium and iron, as well as clay minerals, accumulate
C	Unconsolidated weathered material, little affected by organic activity

Figure 10.6 A schematic soil profile for a well-drained site in a humid-temperate region.

10.4.1 'Ideal' profiles

Traditionally, the typical laterite profile has been described as consisting of three zones: an *iron-rich upper horizon,* resting on a *mottled zone,* with an *iron-depleted pallid zone* at the base, the whole overlying either fresh rock or unleached saprolite. Such a profile is, however, uncommon, and the view that it is typical seems to result from acceptance of the early idea that laterite forms as a result of the upward movement of groundwater. A more recent idea is that the main varieties of laterite have evolved from a common weathering or alteration zone (*see* Fig. 10.7).

10.4.2 Nature of, and processes in, the alteration zone (see Fig. 10.8)

This zone is normally found below the water table and so the weathering processes are hardly affected by climate. It is made up of two different kinds of saprolite.

- *Coarse saprolite* rests on parent rock, from which it is separated by an irregular weathering front. The types of weathered mineral present tend to depend on the degree of water movement.
- *Fine saprolite,* or *lithomarge* (equivalent to the pallid zone of the traditional profile) is developed above. Here weathering is more advanced, as shown by increasing porosity, reduction in rock strength, and complete or partial alteration of most of the parent rock minerals. At the top, both leaching (washing out) and accumulation may occur. Leaching involves the loss of iron, kaolinite and quartz, with possible cavity development and a pale appearance. Kaolinite and goethite (giving a brown colour) may accumulate in the cavities.

Box 10.3 Variation of chemical weathering products and processes on a global scale

Some 14 per cent of the Earth's land surface undergoes mainly mechanical weathering, and about 86 per cent is affected chiefly by chemical processes. This chemical province may be divided into two broad zones, based on the relative mobilities of silica (SiO_2) and alumina (Al_2O_3).

- Zone of *cheluviation* (chelation and eluviation; *see* Box 10.2), in which weathering solutions contain organic acids (carboxylic or phenolic; *see* Box 8.1) and chelation is dominant. Alumina is more mobile than silica, and a silica-rich residue results. Cheluviation dominates in cool regions where the rate of organic breakdown is low, and podzolisation (*see* Box 10.2) is the chief soil-forming process.
- Zone of *soluviation* (solution and eluviation), in which chelating substances are less important and the dominant process is hydrolysis (action of protons) which results in solution. Alumina is less mobile than silica, so desilification (loss of silica) varies, and three contrasting regions can be identified:
  - Region of *bisiallitisation*. The loss of silica is *moderate* here, and so '2:1-layer' clay minerals form (e.g. smectites) with partial retention of basic cations (Na, K, Ca and Mg) in charged interlayer spaces. It occurs in temperate regions and steppes.
  - Region of *monosiallitisation*. The loss of silica is considerable, and '1:1-layer' clay minerals (e.g. kaolinite) tend to form. Basic cations are removed. It occurs in tropical subhumid regions.
  - Region of *allitisation*. The loss of silica is intense, and so aluminium octahedra dominate, giving hydroxides (gibbsite, $Al(OH)_3$) or sometimes oxohydroxides (boehmite, $AlO(OH)$) (*see* Table 10.2). This region coincides with the humid tropics.

The relative importance of these four divisions of the Earth's land surface is shown in Table 10.3 below.

Table 10.3 Proportions of the Earth's 'chemical' environment occupied by the various alteration provinces and main processes

Nature of the alteration division	Surface area (% of chemical zone)	Type of mechanism	Area (%)
Bisiallitisation	45	Soluviation (hydrolysis)	82
Monosiallitisation	21		
Allitisation	16		
Podzolisation	18	Cheluviation	18
Totals	100		100

Source: data derived from Pedro, G. 1968: *Distribution des principaux types d'altération chimique à la surface du globe. Revue de Géographie physique et de Géologie dynamique* **10**(5), 457–70, on which the material in this box is based.

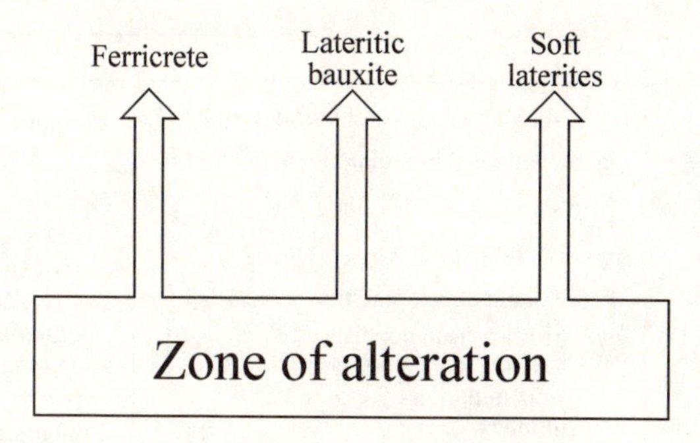

FIGURE 10.7 The main types of laterite may evolve from a common alteration zone.

10.4.3 The ferricrete profile (see Fig. 10.9)

This type of profile is dominated by the transfer and concentration of iron, particularly in the upper zone, where Fe_2O_3 may be as high as 20 per cent. Ferricretes are best developed under a seasonal climate, where average temperatures are about 28°C, rainfall is about 1500 mm per year and the

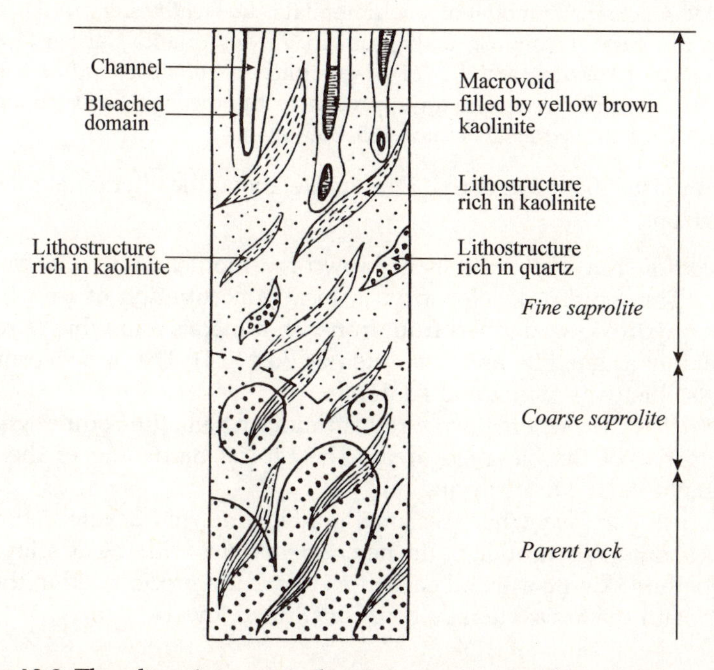

FIGURE 10.8 The alteration zone, showing coarse saprolite, and fine saprolite (lithomarge) with zones of leaching and accumulation. (Based on a diagram in Tardy, Y. 1992: Diversity and terminology of lateritic profiles. In Martini, I.P. and Chesworth, W. (eds) *Weathering, soils and paleosols*. Developments in Earth Surface Processes 2. Amsterdam: Elsevier, 379–405.)

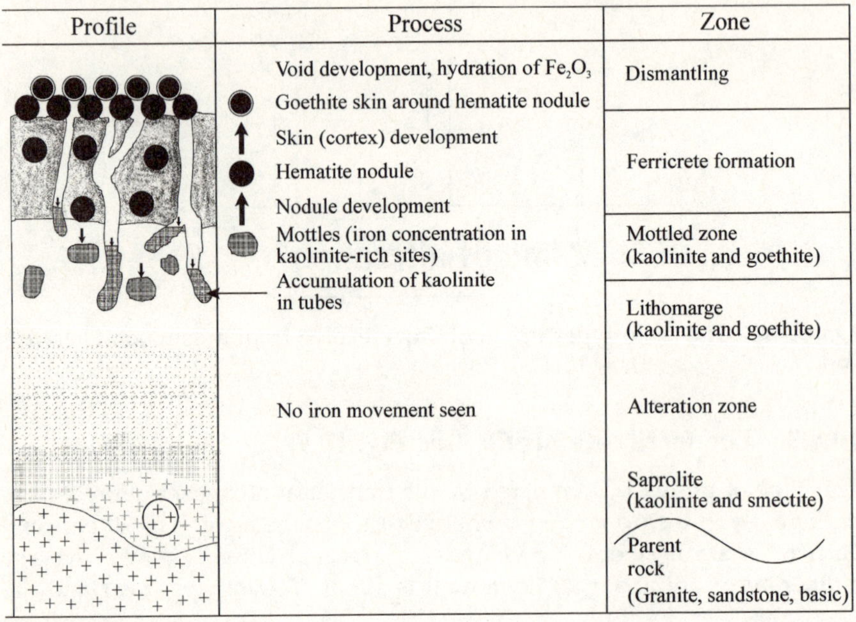

Profile	Process	Zone
	Void development, hydration of Fe_2O_3	Dismantling
	Goethite skin around hematite nodule	
	Skin (cortex) development	Ferricrete formation
	Hematite nodule	
	Nodule development	
	Mottles (iron concentration in kaolinite-rich sites)	Mottled zone (kaolinite and goethite)
	Accumulation of kaolinite in tubes	Lithomarge (kaolinite and goethite)
	No iron movement seen	Alteration zone
		Saprolite (kaolinite and smectite)
		Parent rock (Granite, sandstone, basic)

FIGURE 10.9 A schematic model of a common lateritic weathering profile and its overlying ferricrete. (Based on a diagram in Tardy, Y. and Nahon, D. 1985: Geochemistry of laterites, stability of Al-goethite, Al-hematite, and Fe^{3+}-kaolinite in bauxites and ferricretes: an approach to the mechanism of concretion formation. *American Journal of Science*, **285**, 865–903.)

dry season lasts about 5 months. The profile above the alteration zone has three horizons.

- The *mottled zone*. This consists of brown-red mottles in a white or grey matrix. The mottles develop from a local concentration of iron oxides and oxohydroxides, derived from iron-rich minerals and which precipitate mainly as goethite and hematite (*see* Box 10.4). The bleached matrix consists chiefly of quartz and kaolinite.
- The *ferricrete zone*, where purple-red aluminium hematite nodules grow at the expense of the bleached areas. The typical hardening of the zone appears to be due to its drying out.
- The *dismantling zone*, where seasonal rainfall aids void development and brings about a rehydration of the hematite nodules with the development of a goethite skin ('cortex'). Leached elements are precipitated in the dry season, and the ferricrete may be reconstituted lower down.

10.4.4 Soft laterites

These are sometimes called oxisols, from their oxide content. In many humid tropical areas the regolith of the upper part of the profile is soft and lacks ferricrete development.

Box 10.4 Factors influencing the behaviour of goethite and hematite in weathering solutions

- **The extent to which Al^{3+} is substituted for Fe^{3+}.**
  - Goethite may have less than 33% mole fraction of AlO(OH), giving aluminium goethite. Hematite may have less than 15% mole fraction of Al_2O_3, giving aluminium hematite.
  - Certain minerals may affect the degree of substitution. For goethite, aluminium substitution is 2–20 per cent mole fraction when kaolinite is present, and 18–27 per cent when the gibbsite of bauxite is present. Hematite is also less aluminous when kaolinite and quartz are present.
  - *Generally, aluminium substitution increases mineral disorder and decreases resistance to weathering.*
- **Kinetics (i.e. rates) of precipitation**. Laboratory experiments suggest that, when iron from silicates is released into solution, goethite crystallises relatively early and ferrihydrite (a precursor of hematite) later. Hematite is favoured over goethite, when there is a rapid release of iron and a low concentration of complexing organic compounds. A high concentration of organic compounds favours goethite precipitation.
- **Pore size and water activity**. Solutions saturated with respect to iron minerals may precipitate goethite in water of high activity at the margins of large pores (e.g. in sands), and hematite in water of low activity (low internal pressure) in small pores (e.g. in clays).
- **Differences in Eh value**. Goethite is stable in an environment where the Eh value is +0.887 V. An Eh of +0.2 V would increase its solubility 10 000 times.

In some cases, e.g. in Cameroon, Africa, dark-red nodules of kaolinite and hematite are set in a soft, red, clay matrix that becomes more yellow towards the surface with increasing amounts of goethite. Such a profile seems to be transitional towards a full ferricrete. Elsewhere, soft laterite profiles show a well-drained, iron-rich unit, resting on saprolite and lacking horizon development.

10.4.5 Alumina-enriched laterite (alcrete)

Under suitable conditions an alumina-rich zone may develop in a tropical laterite. If the alumina is sufficiently concentrated for economic working, the formation is called *bauxite*.

Alumina enrichment is due to the local immobility of aluminium and the relatively high rate of silicon loss. This involves 'incongruent' dissolution,

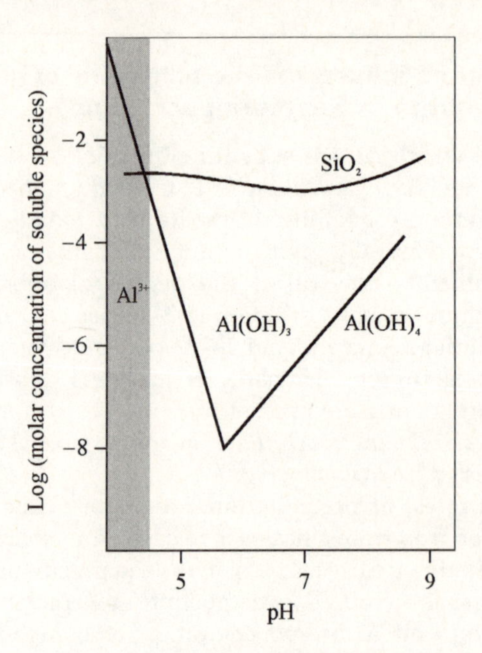

FIGURE 10.10 The solubility of aluminium and silicon as a function of pH. Note that in the shaded zone (pH<4.5), aluminium is more soluble than silicon. This may be the case in the upper part of weathering profiles in humid temperate regions. (*Source*: Raiswell, R.W., Brimblecombe, P., Dent, D.L. and Liss, P.S. 1980: *Environmental Chemistry*. London: Edward Arnold, p. 68.)

i.e. the selective dissolution of a mineral structure. For a weathering feldspar, selective silicon loss occurs at the alteration stages:

$$\text{feldspar} \xrightarrow{\text{loss of Si}} \text{kaolinite} \xrightarrow{\text{loss of Si}} \text{gibbsite } (Al(OH)_3)$$

This contrast in behaviour is largely driven by pH. The solubility of aluminium is lower than that of silicon in the pH range 4.5–9 (*see* Fig. 10.10) which covers most tropical regoliths. High precipitation (>1700 mm per year) and free drainage also favour preferential leaching of silicon. You should note two modifications to this pattern of aluminium enrichment.

- *Variation in the degree of hydration*. Gibbsite ($Al(OH)_3$), which is hydrated, is found in wetter sites. The dehydrated version of gibbsite is boehmite:

$$Al(OH)_3 \text{ (gibbsite)} \rightleftharpoons AlOOH \text{ (boehmite)} + H_2O$$

  This mineral is found in drier conditions.
- *Variation in the relationship between iron and aluminium oxides*. The iron oxide hematite and its hydrated version goethite are usually found with the aluminium oxides. However, the dehydrated hematite tends to occur at the top of gibbsitic profiles. The dehydration of gibbsite to boehmite at

and near the land surface tends to be associated with the removal of iron and its concentration lower down.

10.5 Silcrete

The laterites you have been studying tend to show concentration of iron and aluminium oxides and depletion of silicon. A *silcrete* is a hardened weathering formation showing a marked concentration of silica (silicon dioxide, SiO_2) in horizons sometimes up to 8 m thick. In this section you will be introduced to the chemistry and formation of silcrete.

10.5.1 Silcrete chemistry

This is dominated by silica. Silcretes in south Cape Province, South Africa, show a mean SiO_2 concentration of 94.7 per cent, while samples from South Australia show figures varying between 97 and 99 per cent. The main mineral form is quartz, although opal, consisting of packed spheres of hydrous silica less than 5000 Å in diameter, is also found.

Anatase (titanium dioxide, TiO_2) is often enriched, with a mean concentration of 1.82 per cent for the Cape Province silcretes.

The elements Fe, Al, Ca, K, Mg and P are almost entirely depleted, although the sum of Fe_2O_3 and FeO may reach 0.2–1.5 per cent in some West Australian silcretes.

10.5.2 Conditions and processes of formation

Silcretes tend to develop above, or at the level of, the water table, although both varieties may be found in the same profile. In each case silicon-rich rocks and acid conditions are necessary.

The first type is a *pedogenic* ('soil-forming') *silcrete*. Weakly acid rainwater dissolves quartz-rich surface rocks to form a solution containing up to 6 ppm SiO_2 (the solubility of quartz at ambient temperatures). The solution is concentrated as it percolates through the profile, up to perhaps 120 ppm, and precipitation of quartz in various forms takes place.

The *groundwater silcrete* is developed near the water table, and two varieties occur. The first results when the movement of silica-loaded groundwater is held up at the water table, and precipitation takes place. In the second case, active weathering of suitable rocks at the water table leads to the retention of silicon and titanium ions. Opal is a form of silica precipitation under these conditions.

10.6 Calcrete

Calcrete is a type of chemical sediment that may be locally hardened. It consists of mainly fine-grained, low-magnesium calcium carbonate in the

Box 10.5 Some important minerals in calcretes

Calcite

A common carbonate mineral, having the formula $CaCO_3$. In practice, variable amounts of magnesium may be found, giving low-magnesium (less-soluble) and high-magnesium (more-soluble) calcites tending towards dolomites (Ca, $Mg(CO_3)_2$).

Micrite

A dense, microcrystalline calcite, having a grain size $<4\,\mu m$. It may form within calcrete as a result of precipitation from carbonate-saturated solutions. It may make up 50 per cent or more of a massive calcrete.

Sparite (sparry calcite)

A coarse, crystalline calcite that occupies voids in a developing calcrete. It may result from recrystallisation.

Aragonite

A polymorph of calcite, i.e. it has a different crystalline structure but the same chemical composition. It has no cleavage and a higher specific gravity (2.9). It is less stable than calcite.

form of calcite (*see* Box 10.5) found in the weathering zone above the water table. Its formation is favoured under a climate which is transitional between arid, when too little precipitation only allows superficial accumulation of carbonate, and humid, when soluble minerals are leached away.

In this section you will study the source, precipitation and secondary alteration processes of the calcium carbonate in calcrete.

10.6.1 Sources of calcium carbonate

The following are some of the more important sources.

- Parent rocks rich in calcium carbonate.
- Breakdown of calcium-rich components in the regolith, such as suitable primary minerals, and the shells of organisms.
- Solutes in rain- or ground-water. The average calcium content of rainwater is about 6–7 ppm. Taproots may draw up calcium-rich water over distances up to 50 m.
- Atmospheric dust, from deserts such as the Sahara.
- Marine aerosols in coastal areas.

10.6.2 Mechanisms of calcium carbonate precipitation (see also Chapter 7)

Precipitation is a response to a particular behaviour of the carbon dioxide-water-calcium carbonate system, and to the environmental conditions that encourage that behaviour.

Precipitation and the CO_2–H_2O–$CaCO_3$ system

The reaction involved in this system can be summarised in the general equation:

$$CO_2 + H_2O + CaCO_3 \rightleftharpoons Ca^{2+} + 2HCO_3^-$$

The solubility product $[Ca^{2+}] \cdot [CO_3^{2-}]$ is a measure of the solubility of calcium carbonate and so of the tendency to solution or precipitation. It is affected by independent variables such as temperature, pH and partial pressure of carbon dioxide. Environmental conditions may affect these variables in the direction of precipitation.

Environmental conditions encouraging precipitation

These include

- *evaporation,* which is only important in the upper metre or so of the regolith;
- *increase in soil suction pressure* as plant roots withdraw water;
- *activity of microfloras* (microscopic plants), which can increase pH;
- *plant photosynthesis,* which involves the removal of carbon dioxide;
- *decay of organisms,* which may release ammonia and amines (organic compounds derived from ammonia), thus raising pH.

10.6.3 Secondary alteration processes

Once formed, a calcrete may be modified by further weathering processes, which include the following.

- *Fragmentation,* due to the expansion that may occur when calcite is precipitated in pores of poorly cemented rock. Plant roots may also cause break-up.
- *Conversion of a sparite cement to micrite* (*see* Box 10.5). This may occur through organic action, when the hyphae of algae and fungi (*see* Box 8.5) bore holes in a sparite which are then filled by micrite. Conversion may also occur because of changes in the chemistry of circulating solutions.
- *Recrystallisation.* Microcrystalline calcite may recrystallise to (micro)-sparite along the voids and fractures of layered and massive calcrete.

10.7 Economically valuable weathering residues: metals

Weathering processes may bring about a concentration or refinement of economically useful substances, especially metals. This is particularly the case in humid, tropical regions where lateritic weathering, involving some of the more mobile elements and cations, leads to a concentration of substances that may include those of economic value. In this section we will examine gold, manganese, copper and iron as weathering residues. Bauxite was reviewed earlier in section 10.4.5.

10.7.1 Gold

In its primary or original form, gold occurs mainly in hydrothermal veins (precipitates from hot solutions) connected with acid to intermediate plutonic rocks (igneous rocks formed at great depth). Some volcanic rocks may also contain gold. When gold-bearing rocks are weathered, the metal may be mobilised and ultimately precipitated in a more concentrated form. Several lines of evidence suggest that gold may be dissolved and precipitated under suitable weathering conditions.

- *Changes in the size and shape of gold particles on exposure to weathering*. This process has been described for Gabon, West Africa, where an equational climate with an average annual rainfall of 2000 mm drives active lateritic weathering. At the study site, particles of gold containing 5–9 per cent silver occur in quartz veins in a metabasite (a metamorphic rock), itself enclosed in a granite-gniess (*see* Fig. 10.11). The effect of weathering is to
  - increase particle roundness;
  - decrease the average size of particles, from 1000 μm in the unweathered vein to 500 μm at the top of the saprolite.
- *The locations of gold concentrates*. These tend to be sites where enrichment would be expected if weathering solutions were responsible, i.e.
  - at the water table, for example at a site in Papua New Guinea, where a thin horizon of gold occurs at the present level of the water table in laterised, ultrabasic rocks;
  - in fractures and voids, as in parts of Western Australia;
  - at the surface of weathering products such as iron nodules, concentrations of manganese oxides and weathering clays, for example at a site in Mali.
- *The purity of 'precipitated' gold*. Gold found in the locations described is much purer (i.e. depleted in silver) than in the primary state. At the site described in Fig. 10.11 the silver content of primary gold ranged from 5 to 9 per cent. An extreme case of purity has been described in Kalgoorlie, Western Australia, where particles having a gold content of 99.91 per cent

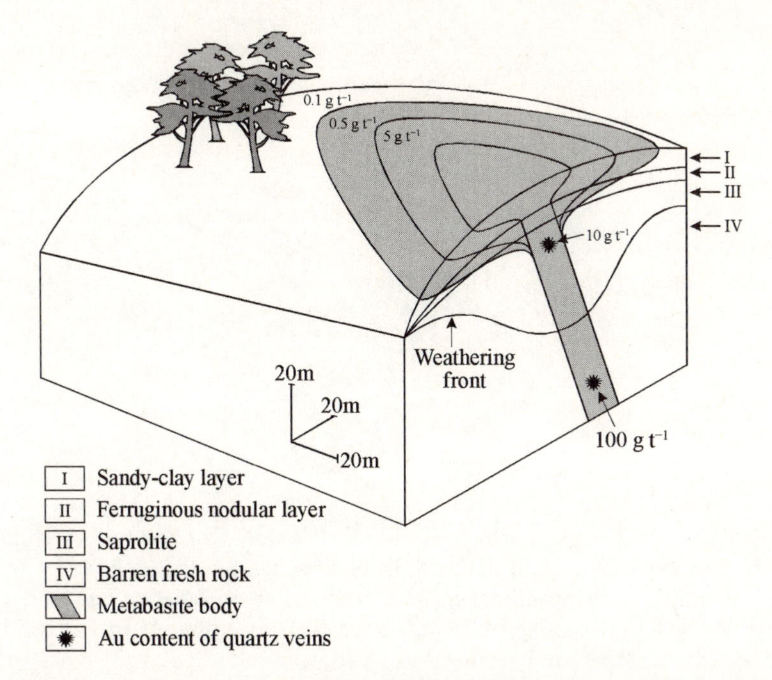

FIGURE 10.11 Weathering horizons and gold content levels ($g\,t^{-1}$) for a locality in Gabon, West Africa. (Based on diagram in Nahon, D.B., Boulangé, B. and Colin, F. 1992: Metallogeny of weathering: an introduction. In Martini, I.P. and Chesworth, W. (eds) *Weathering, soils and paleosols*. Developments in Earth Surface Processes 2. Amsterdam: Elsevier, 445–71.)

have been found. The implication is that weathering acts as a refining agent.

- *The crystalline form* of some gold suggests precipitation.

Investigations suggest that gold may be dissolved by forming a complex with other substances. This may take place under two sets of conditions.

Gold formed in acid, chloride-bearing solutions in the presence of strong oxidant

These conditions are found in the weathering profiles of Western Australia, where chloride ion is provided by rain-bearing winds from the coast but where the climate is generally arid. Chlorine and oxygen react with gold and silicon to form the chloride complex $AuCl_4^-$ and the compound AgCl. The reaction for gold is

$$4Au + 16Cl^- + 3O_2 + 12H^+ \longrightarrow 4AuCl_4^- + 6H_2O$$

and for silver

$$4Ag + 4Cl^- + O_2 + 4H^+ \longrightarrow 4AgCl + 2H_2O$$

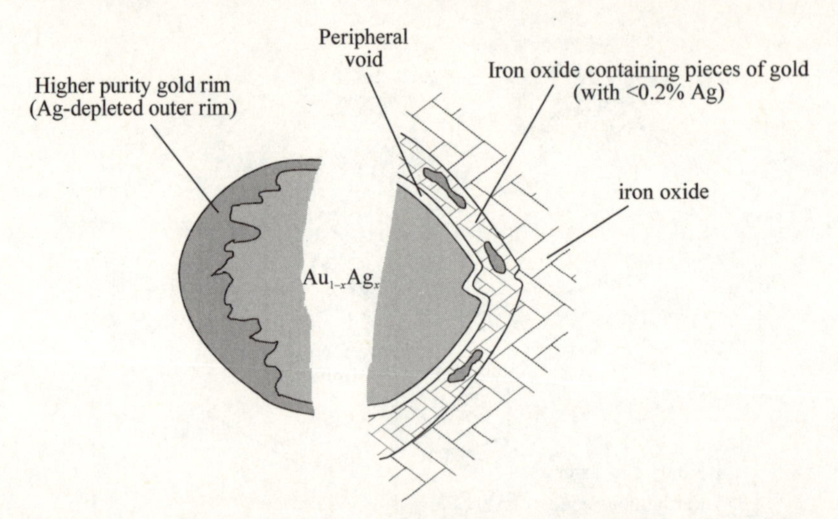

Figure 10.12 Two ways in which a gold particle may respond to its weathering matrix (see text). (Based on diagram in Nahon, D., Boulangé, B. and Colin, F. 1992: Metallogeny in weathering: an introduction. In Martini, I.P. and Chesworth, W. (eds) *Weathering, soils and paleosols*. Developments in Earth Surface Processes 2. Amsterdam: Elsevier, 445–71.)

Iron or manganese ions, acting as reducing agents, then bring about the simultaneous precipitation of both very pure gold and oxohydroxides:

$$AuCl_4^- + 3Fe^{2+} + 6H_2O \longrightarrow Au + 3FeOOH + 4Cl^- + 9H^+$$

$$AuCl_4^- + 3Mn^{2+} + 6H_2O \longrightarrow Au + 3MnOOH + 4Cl^- + 9H^+$$

The situation is different in more humid, tropical environments, where chloride concentrations are less because of dilution. The oxidising agent may be ferric iron (Fe^{3+}), which is less effective than oxygen but which reaches high concentrations at the top of lateritic profiles. In addition, the ligand may be supplied by organic acids. Humic acid, originating through microbial action on decaying organic matter, is present in soil water and can generate a humic acid complex of gold by reaction with traces of the metal. The complex could readily migrate and then precipitate.

Under these circumstances, gold tends to be dissolved and not precipitated at the same time as iron oxides. Consequently, gold is crystallised separately in voids in iron oxide-rich horizons. The dissolution of a gold particle proceeds in one of two directions (Fig. 10.12). Where the particle is in contact with iron oxides, fragments are released into the oxide. Where the contact is with a clay matrix there is a preferential depletion of silver.

Gold formed in alkaline weathering fluids

When rocks containing gold, sulphides and carbonates are oxidised, alkaline weathering solutions tend to result. Such solutions are unsuitable for the transport of gold as a chloride complex.

However, they may be suitable for the formation of gold complexed by the thiosulphate ion $S_2O_3^{2-}$; ('thio' means the replacement of oxygen by sulphur). The problem is that thiosulphate has only a limited stability. Its most favourable environment for stability is in the phreatic zone (i.e. below the water table).

The nature and mineral associations of secondary gold in the oxidised zone of carbonate ore at Wau, Papua New Guinea, is consistent with earlier remobilisation as a thiosulphate complex. Here the gold is coarsely crystalline, alloyed with 50–75 per cent silver and enriched at the water table and with MnO_2 in the oxidised zone.

10.7.2 Manganese

The development of economic concentrations of manganese involves the oxidation of, for example, silicate or carbonate phases of the metal (as Mn^{2+}) into oxohydroxides with Mn^{3+} and then into oxohydroxides and oxides with Mn^{4+}. The nature of both the parent rock and the weathering solutions affects the process.

- Different parent minerals can generate the same oxidised weathering phase. For example, both rhodochrosite (a carbonate, $MnCO_3$) and rhodonite (a pyroxenoid, a type of silicate, $MnSiO_3$) weather ultimately to pyrolusite (MnO_2). In this case the nature of the weathering solutions determines the nature of the secondary phase.
- Under other conditions the same mineral may yield different secondary products.
  - Under high pressures of carbon dioxide, rhodochrosite weathers directly into birnessite ($(Na,Ca)Mn_7O_{14} \cdot 3H_2O$), but under lower carbon dioxide pressures it weathers into manganite ($MnO \cdot OH$).
  - A complication is added if potassium is available in the weathering environment, especially when potassium feldspars, muscovites or illites are weathered at the same time. In this case rhodochrosite alters into the oxide cryptomelane (KMn_8O_{16}).

As weathering proceeds, porosity increases, more atmospheric oxygen becomes available and the oxidation state of the manganese increases, i.e. oxidation occurs. Locally, a very oxidised phase may grade into a less oxidised phase. The reason for this is that certain primary minerals are slow to weather and thus to release reducing ions which reverse the oxidation status of the manganese oxide. For example, the late release of potassium from mica-rich minerals indirectly brings about the conversion of pyrolusite ($Mn^{4+}O_2$) into the less-oxidised cryptomelane (KMn_8O_{16}).

The role of aluminium is important in the strongly weathered upper horizon. Here the relative concentration of manganese oxides can locally produce very acidic environments, which allow the release of aluminium from kaolinite. This metal is then incorporated in the developing lithiophorite ($(Al,Li)MnO_2(OH)_2$), which may be produced in large amounts.

However, further alteration may lead to the breakdown of lithiophorite and the release of aluminium with the formation of horizons rich in aluminium hydroxides. This situation confirms that, in lateritic environments, manganese is more mobile than aluminium.

10.7.3 Copper

Copper enrichment in weathering horizons is strongly influenced by two factors:

- the high mobility of copper as an element;
- the variety of parent materials, and particularly the extent to which they may be fractured.

It is not, therefore, realistic to think in terms of a typical copper weathering profile. A more useful approach is to study the sequence of copper-bearing secondary minerals that develop from the parent rock as weathering proceeds. The resulting copper concentrations have been studied in Zaire and Brazil.

The source of copper in weathering profiles in these countries consists of primary sulphides, mainly chalcopyrite ($CuFeS_2$), which occurs in lenses or veins in the host rock, and secondary sulphides, such as bornite (Cu_5FeS_4), which are the result of alteration by groundwater.

The weathering sequence proceeds with the development of carbonate phases, particularly malachite ($Cu_2CO_3(OH)_2$) which is relatively stable. Maturity develops with the emergence of copper silicates such as dioptase ($CuSiO_3 \cdot H_2O$) and chrysocolla ($Cu_8Si_8O_{20}(OH)_8 \cdot 8H_2O$). At this stage the quartz and silicates of the host rocks are dissolved, with most of the silica being mobilised.

The replacement of quartz by dioptase is common in the weathered deposits of Mindouli in Zaire, while its replacement by chrysocolla is characteristic of the profile of Santa Blandina in Brazil. The end product of the mineral weathering sequence is a strongly oxidised zone, when cuprite (Cu_2O) is found.

The high mobility of copper means that it is readily taken up by percolating solutions. It may then be precipitated as copper-bearing carbonates and silicates in the weathering products of host rocks. It follows that zones of copper enrichment do not necessarily lie above the parent copper sulphides.

10.7.4 Iron

Iron may be oxidised and reduced (*see* Chapter 7) in surface and near-surface environments. Oxidation is the important process as far as the retention of secondary iron is concerned, and is associated with the development of iron-rich horizons which may be sufficiently concentrated for economic exploitation.

In deep-seated rocks iron is often in the ferrous (divalent) state. This is the case in primary minerals such as magnetite ($Fe^{2+}(Fe^{3+})_2O_4$), siderite ($FeCO_3$), pyrites (FeS_2), and the ferromagnesian silicate minerals. The ferrous iron is released, typically by acid attack by carbon dioxide, and may then stay in solution or be oxidised and precipitated according to the environmental conditions.

For ferrous iron to remain soluble it requires acid conditions (pH <7) and a reducing environment with organic matter present. Under these conditions, the solubility of iron may reach 50 mg dm^{-3}. However, the more usual path is for ions of ferrous iron to come into contact with dissolved oxygen, whereupon they are oxidised and precipitated as ferric oxide. In this state, and in the presence of oxygen, iron is practically insoluble; its maximum solubility is about 1 mg m^{-3}.

With iron-rich parent rocks, under suitable environmental conditions, and after a long period of time, the near-surface zone is leached of mobile substances and enriched in ferric ion.

READINGS FOR FURTHER STUDY

A most useful book, containing chapters on alteration products (especially the clay minerals), laterites, silcretes and calcretes, and the concentration of metals by weathering, is

Martini, I.P. and Chesworth, W.P. (eds) 1992: *Weathering, soils and paleosols.* Developments in Earth Surface Processes 2. Amsterdam: Elsevier.

You will find that many of the papers referenced in this chapter elaborate upon the ideas we have outlined. Here are some more articles that you should also find helpful.

Arakel, A.V. 1982: Genesis of calcrete in Quaternary soil profiles, Hutt and Leeman lagoons, Western Australia. *Journal of Sedimentary Petrology* **52**, 109–25. This case study contains useful material on the nature and origin of calcretes.

Bourman, R.P. 1993: Perennial problems in the study of laterite: a review. *Australian Journal of Earth Sciences* **40**, 387–401. An examination of the terminology and classification of laterite, its mode of formation, age and use in the reconstruction of past climates.

McFarlane, M.J. and Bowden, J.B. 1992: Mobilisation of aluminium in the weathering profiles of the African surface in Malawi. *Earth Surface Processes and Landforms* **17**, 789–805. Provides evidence for the view that iron-rich laterites develop when congruent dissolution of kaolinite occurs, i.e. silicon and aluminium are simultaneously removed. Aluminium may be mobilised in an organically bound form. You will also find that this article will help your understanding of other topics, such as etchplain development (*see* section 11.4.3) and the value of aluminium as an index mineral in studies of weathering intensity.

Thiry, M. and Millot, G. 1987: Mineralogical forms of silica and their sequence of formation in silcretes. *Journal of Sedimentary Petrology* **57**, 343–52. Describes the origin and order of formation of varieties of silica found in some French silcretes. The sequence is partly controlled by the solubility of each component.

Webster, J.G. 1986: The solubility of gold and silver in the system Au–Ag–S–O_2–H_2O at 25° and 1 atm. *Geochemica et Cosmochimica Acta* **50**, 1837–45. This paper develops the view that the solubility of gold (and silver) in other than acid solutions may be due to complex formation by sulphur-bearing ligands. Precipitation of gold occurs when the combined gold–silver complex breaks down.

11

WEATHERING-CONTROLLED LANDFORMS

11.1 General principles

Weathering-controlled landforms are those whose formation is dominated by weathering processes. Erosional, depositional and complex varieties may be identified (*see* Fig. 11.1), and some of the main types are reviewed below.

Erosional types are formed when weathering selectively attacks bedrock. The removal of weathered debris often, however, involves transportation processes such as creep or the mechanical forces exerted by running water. Such processes are not normally seen as part of the weathering process. If the transportation component is, however, intimately linked to the weathering process itself, then an erosional, weathering-generated landform may be recognised. An example of a linked transportation process would be the removal in solute form of bedrock. The removal of material by the action of lichen and algae would provide another example.

A further relief-forming action of weathering is to produce *depositional features*. In such cases the landforms are the result of linked transportational processes acting on the results of weathering. An example is the scree which builds up when rock debris is weathered from a cliff face. The forms of chemical accumulation in limestone caves provide further illustrations of this type.

The final category of weathering-generated landforms includes those termed *complex*. In this type weathering plays a major role, but unrelated transport processes (cf. the erosional, weathering-generated type above) actually generate the relief. At the smaller scale the tor, fashioned by selective weathering but exposed by erosion, is an example of this category. Landforms underlain by duricrusts in tropical and subtropical environments provide larger examples. Weathering in these environments has given rise to large-scale 'neoformations' which, when they consist of hardened horizons,

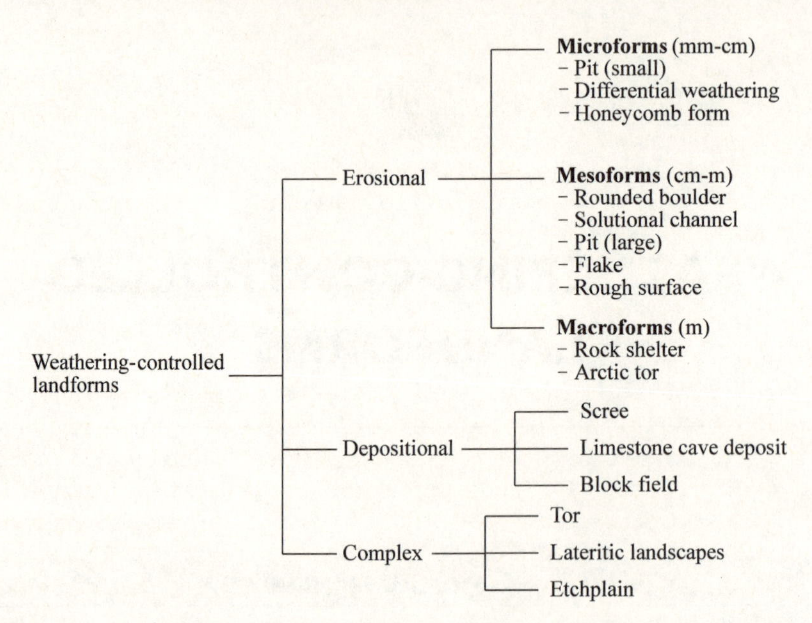

FIGURE 11.1 A classification of weathering-controlled landforms based on primary origin and, in the case of the erosional group, size.

e.g. ferricretes, overlying softer material, may give rise to distinct landforms when they are selectively eroded. The etchplain, an assemblage of landforms dominated by the relationship between weathering and erosion, provides the largest illustration in this category.

11.2 EROSIONAL LANDFORMS

11.2.1 Microforms

You should appreciate that many of the smaller erosional landforms cannot easily be attributed to a single process. Chemical alteration may exploit an existing weakness, perhaps a set of microcracks, and may be associated with biochemical activity, often due to lichen colonisation. Under suitable conditions, salt crystallisation may also act to exploit weaknesses.

A small *pit* is a rounded hole having a diameter of a few millimetres up to about 1 cm and a similar depth. Detailed studies show that a hollow begins by the dissolution of dark-coloured minerals and is rounded by the removal of small grains and mineral fragments.

Small-scale *differential weathering* forms develop when materials of different composition occur next to each other. In a layered metamorphic rock such as gneiss there may be marked depressions along bands that contain mica, hornblende and other mafic (i.e. containing magnesium and iron) minerals. Veins and nodules of resistant quartz often stand well above the adjacent surface, perhaps by 2 cm. The contrast in degree of weathering is often enhanced by the mechanical and biochemical action of lichens.

The *honeycomb* form consists of a group of holes each a few centimetres deep and of similar diameter, with a tendency to widen inwards. The holes tend to occur at sites of microcrack development, and their growth is encouraged by lichen activity. At coastal locations, salt crystallisation may contribute to their development.

11.2.2 Mesoforms

A *rounded boulder* (or corestone) is formed by spheroidal (acting on all sides) weathering. This mesoform is typically developed in igneous rocks, including granites and basalts. The stages in the emergence of a rounded boulder have often been described. Initially square-shaped, joint-bounded blocks develop, with grus (platy fragments) in the cracks. In later stages, well-rounded boulders appear. Several ideas have been proposed to explain the rounding process. One theory is based on the idea that the pattern of weathering is strongly influenced by the arrangement of earlier microcracks, especially when these have a rectangular arrangement. Other ideas involve chemically induced expansion, pressure release, and the development of alternately enriched and depleted weathering layers, called *Liesegang rings*.

A *solutional channel* is a common feature on an exposed sloping surface of a soluble rock such as a hard limestone. Where several channels are closely spaced and run roughly parallel to each other, the group term *rillenkarren* is used. Each channel may be about 50 cm long and 1–2 cm wide and deep. Dimensions vary with the environment. Traditionally, such channels were thought to be formed by the action of acid rainwater running down rock surfaces. The results of recent experimental work suggest that biological weathering by algae may also be important. Similar channels, sometimes called 'flutes', may be found on less-soluble rocks such as granites and sandstones. However, they are most typical of karst landscapes (*see* section 7.11).

A *large weathering pit*, up to a metre or so in diameter and several centimetres in depth, is found on an exposed level surface. In temperate regions this form starts as a small depression which periodically fills with precipitation and organic debris. On granitic rocks the resulting acid solution brings about a steady enlargement. On quartz-rich sandstone (e.g. at Fontainebleau, France) iron may be dissolved from the litter and precipitated. This encourages lichen growth, with mineral destruction through the activity of hyphae (*see* Chapter 8). Organic compounds achieve biochemical weathering with the products being removed in solution. Elsewhere, under suitable conditions, salt crystallisation may assist in the enlargement of weathering pits.

A *flake* is a sheet of fairly uniform thickness, perhaps 1–10 mm, which may occupy an area of up to several square metres. The flake may bulge away from the underlying rock, leaving a gap of perhaps a few millimetres. A flake typically shows considerable microcrack development and mineral alteration. It is developed on a range of rock types, including granites, gneisses, sandstones and limestones. Flaking has been described in all environments, and has been attributed to a range of processes, all of which involve expansion of the surface layer. The processes include chemical weathering (especially hydration), insolation, pressure release, salt crystallisation and natural fires.

A *rough surface* is an area of several square metres consisting of individual depressions (1–5 cm deep) that have almost merged. Such a surface is often found on a gneiss, where the hollows develop at mica-rich sites, and at moist locations which favour chemical weathering. The typical presence of microcracks encourages chemical alteration.

11.2.3 Macroforms

Enhanced weathering at the base of a rock outcrop may bring about a relatively rapid and local retreat of the face with the development of an overhang, or *rock shelter*. A normal requirement is variation in rock properties, thus allowing selective weathering. This is the case, for example, in northern England, where porous and well-jointed biomicrites frequently give rise to rock shelters. A classic example of this feature is, however, provided by the western part of the double tor of Haytor Rocks, Dartmoor, southwest England.

Most of Haytor is made of the so-called 'giant' or 'tor' granite, which has a massive structure with well-developed vertical joints about 4 m apart and near-horizontal false-bedding planes about 1.5 m apart. The lower part of the face, where the shelter is developed, coincides with the outcrop of the so-called 'blue' granite. This is divided by several sets of closely spaced vertical joints. The weathering process that produced the Haytor rock shelter was frost shattering, which readily exploited the closely jointed blue granite, while the massively jointed tor granite was relatively immune. The size of the shelter 'roof', or overhang, was limited by the horizontal dimensions of the joint-bounded blocks of the tor granite. It is unlikely that differential chemical weathering formed the shelter as the blue granite, in spite of its closely developed jointing, is relatively resistant to such attack.

11.3 Depositional landforms

11.3.1 Scree

This weathering landform, which consists of a deposit of angular rock fragments resting against a steep slope, is a product of frost shattering and gravitational forces. Screes are being actively formed in environments where frost shattering is an important process, but interesting 'fossil' variants are found in parts of Snowdonia, North Wales, and will be reviewed as an example of this landform category.

In a typical example at Coed Camlyn, Maentwrog, Gwynedd, Wales, the surface of the scree slopes at about 25°. The internal fabric consists of angular shale fragments oriented parallel both to each other and to the ground slope. Where screes were formed from more massive rock types such as rhyolites, the rock fragments are much less platy and so less well oriented. There is little if any sand–silt–clay-sized material within the body of the scree; this is likely to have been washed away as rapidly as it was produced.

The formation of this and similar 'fossil' screes in Snowdonia was encouraged by two factors.

- *The generally cleaved and fissile nature of the shales and slates of the region.* This means that they are readily penetrated by water and so are susceptible to frost weathering.
- *The likely occurrence of climatic conditions favouring such frost shattering.* This took place both at the end of the last glacial phase, when bare rock surfaces were being exposed as the glaciers retreated, and later, during the 'Loch Lomond' stadial (cold phase), 11 000–10 000 BP (years before present).

The fossil nature of many of the Snowdonian screes is indicated by their partial burial beneath a veneer of later soil. Locally this has been removed by erosion, as at Sychnant, and Marian Rhaiadr Fawr (Aber).

11.4 Complex landforms

11.4.1 The tor

A tor is an exposed mass of blocky or rounded rocks, generally undisturbed, and is widely found on massively jointed crystalline rocks such as granites. It also occurs on similarly jointed sediments such as sandstones. It is a complex landform in that weathering plays an important role in its formation, but rock properties and erosional processes contribute significantly. The Dartmoor tors provide classic examples of this landform.

Nature of the Dartmoor granite

The rock properties at each of the three main tor positions in the Dartmoor landscape are shown in Table 11.1. An analysis of granite (per cent) from a specific tor, Saddle Tor, shows:

Quartz	33.7
Potassium feldspar	31.7
Plagioclase feldspar	21.2
Biotite	9.8
Muscovite	0.7
Accessory minerals	2.8

Apart from quartz, these minerals are vulnerable to chemical breakdown.

TABLE 11.1 Some rock properties of Dartmoor tors at the three main locations

Rock properties	Tor position		
	Summit	**Valley side**	**Spur**
Feldspar content	Abundant, >30% potash and >18% plagioclase	Low in potash feldspar (<31%)	
Texture	Strong megacryst development, usually >15%	Feebly megacrystic to equigranular	Feebly megacrystic (<5% megacrysts) to equigranular
Grain size	Coarse	Fine to intermediate	Fine
Joint characteristics	Primary vertical joints widely spaced (>300 cm)	Primary vertical spacings narrow (<300 cm)	Primary vertical spacing <200 cm. Secondary vertical spacing 50–75 cm
	Primary horizontal spacing 60–80 cm. Secondary horizontal spacing >100 cm	Primary horizontal spacing 60–200 cm. Secondary horizontal spacing <10 cm	Primary horizontal spacing <60 cm. Secondary horizontal spacing <10 cm

(*Source*: Ehlen, J. 1992: Analysis of spatial relationships among geomorphic, petrographic and structural characteristics of the Dartmoor tors. *Earth Surface Processes and Landforms* **17**, 53–67.)

Joint character is particularly important, and for Dartmoor as a whole two main sets of joints have been recognised:

- *horizontal joints*, or *pseudo-bedding planes*, which generally follow the slope of the land surface, and may have originated as rock pressure was released due to erosion;
- *vertical joints* show a slight preference for north–south and east–west trends, and may be due to stresses associated with the original emplacement of the granite.

Contribution of jointing to tor form and development

The contribution of jointing to *tor shape* is clear.

- The dome-like shape of Blackingstone Rock, Dartmoor, reflects the dominance of pseudo-bedding planes that separate thin sheets of granite dipping away from the tor summit. Vertical joints seem to be poorly developed.
- By contrast, the form of Hound Tor reflects the similar development of both vertical and horizontal joints. Each of the tor's two main components (there is a central 'avenue') consists of a stack of joint-bounded blocks.

The character and development of jointing underlie an important model of *tor development*.

1. The future landscape may have begun as an etchplain (*see* section 11.4.3) during mid-Tertiary times.
2. A developing drainage pattern would have been controlled by the vertical joint sets, with which there is a good correlation.
3. As the land surface was lowered, some vertical expansion of rock would occur due to pressure release, and horizontal joints would develop. Domes and ridges would emerge, defined by the evolving drainage pattern.
4. Summit tors (e.g. Great Mis Tor, Greater Staple Tor and Haytor) developed at the crowns of domes and ridges. Further incision and release of pressure led to further joint development, and ultimately the formation of avenues by erosion.

Joint character must also have influenced the course of weathering on Dartmoor.

Contribution of weathering

Both physical and chemical weathering processes have been suggested to account for the detailed circumstances of tor emergence.

- *Frost action,* guided by joint density (*see* section 11.2.3), is the main physical process proposed. The theory is that tors are residual features, left behind when surrounding bedrock was shattered by freeze–thaw processes and then removed by mass movement. The presence of clitter (spreads of angular rock debris) around many tors is good evidence for the process. The view that growan (*see* Box 11.1) is a result of mechanical disintegration is less convincing. Many valley-side tors may, however, have been modified if not formed by frost action.
- *Chemical weathering* provides an alternative explanation. The basic types of decomposed granite are reviewed in Box 11.1. There is a clear distinction between the intensely altered, fine-grained, feldspar-poor kaolinite and the coarser, less-decomposed growan. There is debate over the origin of the former; both intense tropical-style weathering, and the action of hot, circulating fluids, have been proposed as the decomposing agencies. The growan is, however, generally seen as a product of weathering, perhaps under conditions similar to those that produced the sandy regolith called 'arène' (*see* Chapter 9). Its presence suggests that the tors should be seen as chemically resistant landforms. In the case of summit tors, such resistance may be due to the strongly megacrystic textures (i.e. consisting of large crystals; *see* Table 11.1) of the underlying rocks, the presence of schorl (tourmaline), and the large amounts of quartz.These properties may be reinforced by vertical joint spacing, which is wider than for any other tor locality.

Tors may then have emerged as major components of an incised etchplain which is still currently evolving. Studies of contemporary weathering processes suggest that the properties of the growan (high feldspar/quartz

Box 11.1 The decomposed granite of Dartmoor

'Kaolinised' granite

When intense alteration has occurred kaolinite results, largely derived from decomposed feldspars. This fine-grained, friable material is mined for china clay. Textures vary, reflecting differences in the parent granite and in the intensity of kaolinisation. Pale brown or white micas, including sericite, which is fine-grained, and illite, are present. Mica has a dioctahedral structure as described in Chapter 10. Where alteration is less intense, orthoclase feldspar remains. Quartz is the major primary mineral inherited from granite.

The alteration may be due to the following.

- *Intense chemical weathering*. This view is supported by the calculated isotopic composition (deuterium/hydrogen and $^{18}O/^{16}O$ ratios) for the near-surface solutions that were active during kaolinisation. They are similar to those observed today in warm temperate to tropical climates.
- *The action of hot circulating solutions* following granite emplacement.

Growan

This is a coarser material. Several lines of evidence suggest a moderate degree of weathering: the feldspar/quartz ratio, an index of weathering intensity, which has an average value of about 1 (cf. 0.13 for intensely altered kaolinite); the hard and shiny nature of many feldspar fragments; the retention of the original texture of the granite, but this is usually easily broken.

This material appears similar to 'arènes' (*see* Chapter 9). Growan may sometimes show evidence of downslope movement ('bedded growan') and may be overlain by a veneer (<2 m thick, as at Bellever Quarry) of blocky debris in a matrix of gritty to sandy loam. This is *'head'*, produced under cold-climate conditions.

ratio; presence of kaolinite and local gibbsite) are consistent with present-day decomposition, and that there is little need to infer past hotter and more humid climates.

11.4.2 Lateritic landscapes

Lateritic mantles, between 10 and 150 m thick and consisting of highly weathered material rich in secondary forms of iron and/or aluminium and depleted in silica, are found widely in tropical environments (30°N–30°S). Their thickness and character, particularly the development of upper hardened horizons, or *duricrusts*, e.g. ferricrete, are such that they provide the 'parent material' for landscape development. Landforms emerge as a consequence of the etching effect of erosion, but the major role of weathering products makes it reasonable to class the associated landforms as examples of the 'complex' variety.

The study of lateritic landscapes has stimulated interest in three problems relating to their evolution.

- Can the nature of the pre-lateritic landscape and its subsequent history be worked out and are such reconstructions realistic?
- How may a laterised landscape evolve?
- What landforms are directly influenced by the character of lateritic profiles?

The problem of the 'pre-lateritic' landscape and its history

The traditional view has been that an already existing peneplain (i.e. a landscape of low relief, cutting across a range of rock types and structures) was necessary for the laterisation process to proceed. By implication, such a surface could be deduced from the presence of laterite. In practice, and because of subsequent erosion, such a surface is usually 'reconstructed' by extrapolation from small existing, laterised fragments. This 'peneplain' interpretation has been adopted at both continental and regional scales.

A continental example: western and coastal Africa

Here laterite-mantled surfaces are extensively developed, and the associated weathering processes may have been active for 100 million years. Landform stability over this period of time is unlikely, and one model proposes a sequence of seven planation surfaces of differing heights and each supporting a laterite of distinctive character (*see* Fig. 11.2).

The upper three surfaces have been interpreted as bauxitic surfaces of Jurassic Cretaceous and Eocene age respectively. The lower four have been seen as ferricrete-capped features. On one view their ages range between Pliocene and Middle Quaternary; on another the ferricrete-mantled surfaces are between 1 and 40 million years old. This sequence of peneplanation surfaces has been attributed to a succession of alternating humid and arid periods which are believed to have operated since Jurassic times.

A regional example: the Sydney district, New South Wales

Layers of hard material formed by natural cements and found at or near a land surface are often called *duricrusts*. A variety is ferricrete, where the cementing material is iron oxide. This has been described for the Sydney area of Australia. One interpretation regards its development as the result of the deep weathering of an existing surface of low relief, possibly close to sea level. Differential uplift then occurred, with the subsequent dissection and widespread destruction of the duricrust.

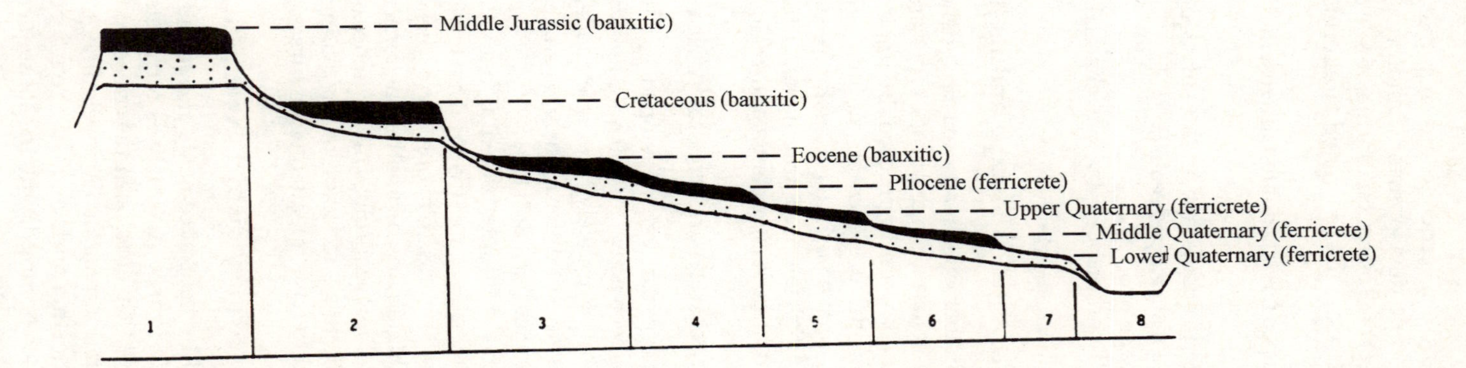

FIGURE 11.2 Seven peneplanation surfaces capped by bauxites or ferricretes and developed between the Fouta Djalon in Guinea and the Niger alluvial plain in Mali (after P. Michel, 1973). (Source: Tardy, Y. and Rouquin, C. 1992: Geochemistry and evolution of lateritic landscapes. In Martini, I.P. and Chesworth, W.P. (eds) *Weathering, soils, and palaeosols*. Developments in Earth Surface Processes 2. Amsterdam: Elsevier, 407–43.)

Challenges to the traditional views

The integration of lateritic fragments rests on the assumption that the separate units have the same characteristics; for example, they are rich in aluminium oxides for the higher African examples referred to, and so demonstrably once formed part of the same horizon, whose age was the same everywhere.

Africa

In the African cases it has been shown that, far from being uniform, separate fragments of the same assumed horizon show definite chemical and mineralogical variety. Their integration would therefore be unconvincing. In addition, detailed studies at a local scale (10 km) in south Mali have shown great variations in kaolinite and hematite content in a continuous ferricrete sheet.

Acceptance of the idea of lateral diversity means that the simplest explanation for the origin of duricrusted landscapes could involve the recognition of a single laterised surface; several are unnecessary. There are several ways of explaining the observed diversity within the idea of a single surface (*see* Fig. 11.3):

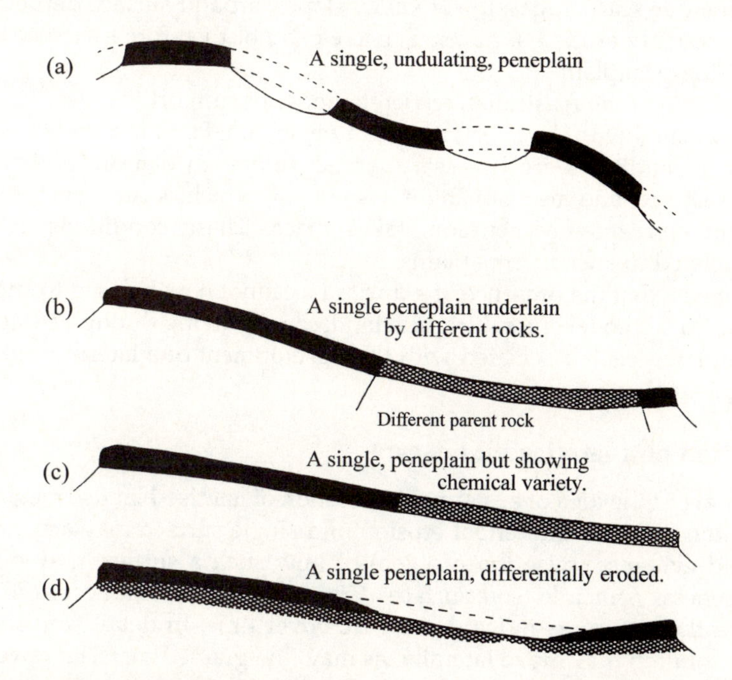

FIGURE 11.3 Some explanations for lateral diversity, on the assumption of a single, basic, lateritic surface. (Source: Tardy, Y. and Rouquin, C. 1992: Geochemistry and evolution of lateritic landscapes. In Martini, I.P. and Chesworth, W.P. (eds) *Weathering, soils, and palaeosols*. Developments in Earth Surface Processes 2. Amsterdam: Elsevier, 407–43.)

- when this involves height differences, they may be accounted for by the erosional dissection of an original undulating surface to produce remaining fragments at different heights (*see* Fig. 11.3(a));
- different parent materials when weathered could give different laterites on the same peneplain (*see* Fig. 11.3(b));
- variation in the degree of lateral movement of weathering solutions could give chemical diversity (*see* Fig. 11.3(c));
- the differential erosion of a single laterised surface could expose different horizons in the regolith and so give rise to lateral diversity (*see* Fig. 11.3(d)).

Australia

A similar argument applies to areas of Australia, particularly the Cape York Peninsula region and the western and south eastern parts of the country. Here the argument that landscape development has passed through a series of stages, each characterised by a peneplain supporting a ferricrete, is not supported by the actual position or likely origin of the ferruginous duricrust.

The positional argument is illustrated in the Cape York Peninsula. Here ferricrete is not limited to high plateaux but occurs in many other locations. These include scarp edges, lower valley slopes, around surface depressions and on recently eroded surfaces. Ferricrete is not therefore evidence of an underlying peneplain.

The origin of the Australian ferricrete does not support the view that it is only associated with the laterisation of a peneplain. For a ferricrete to form, two basic conditions are necessary: an accumulation of iron, perhaps by lateral seepage, and an oxidising environment, which occurs when drying out, and consequently aeration, takes place. These conditions are not solely related to ancient peneplains.

It appears that the presence of a duricrust cannot now be used to support the traditional model of a subtropical landscape evolving through a staircase of peneplains, each associated with the development of a laterised horizon during a stable phase.

Evolution of a laterised landscape

Doubts about the idea of a 'stepped' evolution of laterised landscapes, and a realisation that the differential erosion of a single surface may account for the field evidence, raise the problem of how such a surface may evolve. As a general principle, soil surfaces, iron and aluminium accumulations and weathering fronts move downwards over time. In detail, and locally, iron in solution may move laterally, as may fine-grained material reworked by termites, but the general theme is of vertical evolution.

This vertical development means that laterites are rock controlled, or *lithodependent*, with respect to their chemistry and mineralogy. This lithodependence is not, however, on the rocks below but on those that originally existed above the present land surface. A consequence is that resistant minerals such as gold and quartz may appear dispersed within the contemporary regolith, while their original occurrence may have been concentrated in a vein dipping at an angle with respect to the evolving surface. These ideas of landscape evolution are illustrated in Fig. 11.4.

You should appreciate that the idea that general models of landscape evolution will explain every local circumstance is not very realistic. For

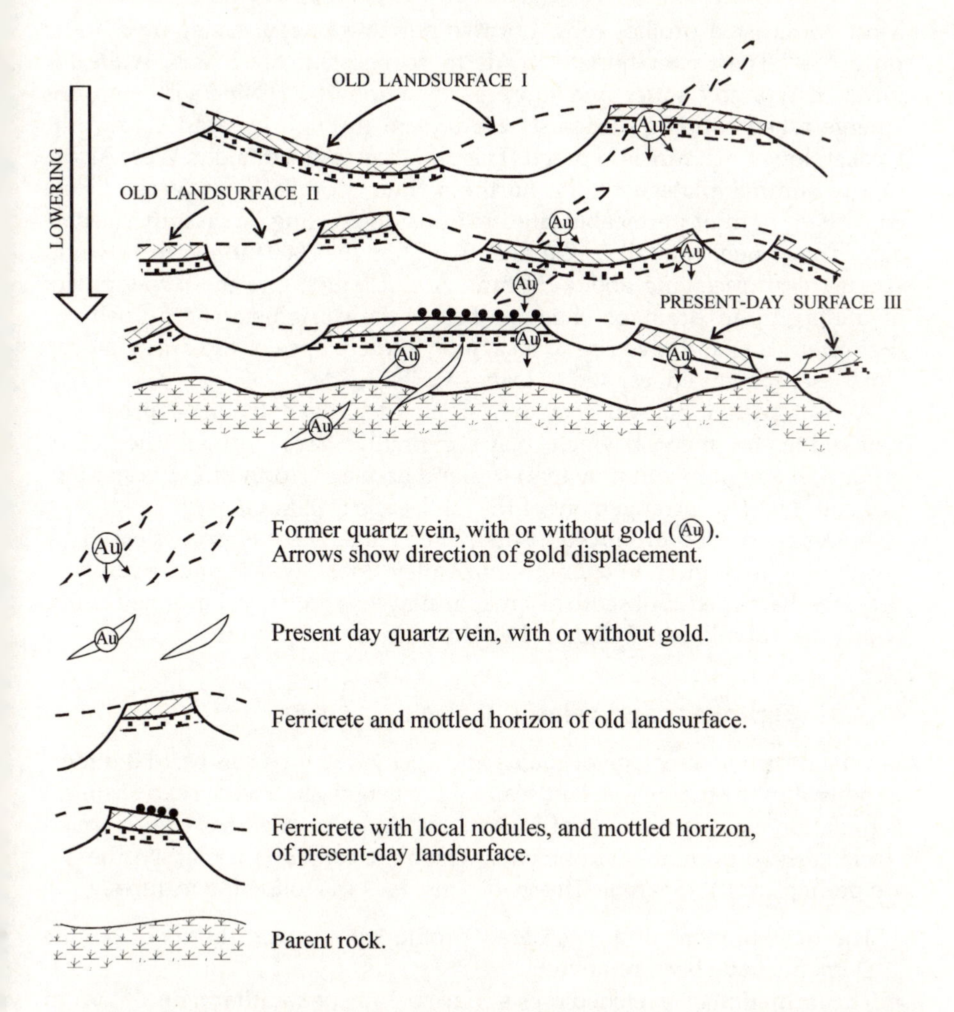

Note that the dip angle of the gold-bearing quartz veins means that the gold is apparently laterally dispersed below the present land surface.

FIGURE 11.4 An outline of the vertical lowering of a lateritic landscape. Note that the dip angle of the gold-bearing quartz vein means that the gold is apparently laterally dispersed below the present land surface. (Based on a figure in: Tardy, Y. and Rouquin, C. 1992: Geochemistry and evolution of lateritic landscapes. In Martini, I.P. and Chesworth, W.P. (eds) *Weathering, soils, and palaeosols*. Developments in Earth Surface Processes 2. Amsterdam: Elsevier, 407–43.)

example, the bauxite-capped Birrimian ranges of central Ghana show an extensive, nearly level summit surface, fringed by steep escarpments 300 m in height. There is no evidence here of vertical movement of the surface duricrust, so the 'classical' model does not apply.

Landforms reflecting the nature of the lateritic profile

Most duricrusted profiles consist of two horizons of contrasting mechanical properties. The upper horizon tends to be resistant to erosion, while the lower is typically softer and more easily removed. Distinctive landforms emerge when capping duricrusts are broken through by erosion, and the weaker, lower material is exposed. This situation is illustrated in West Africa.

The summit plateau of the northern Sula Mountains in Sierra Leone consists of 12 m of permeable lateritic ironstone resting on an impermeable clay. The annual rainfall of the area is high, over 3000 mm, with July and August each receiving about 750 mm. A consequence is the development of underground drainage channels below the duricrust. These flush out the softer clays and give rise to a sagging and collapse of the duricrust cap along escarpment edges, and to extensive landslides.

A duricrusted plateau may be largely destroyed by such processes. Remaining fragments, which may originally have formed the lower ground in a region, will now tend to stand above surrounding, present-day low country. This arrangement of the landscape is *inverted relief.*

Inversion can be produced at a smaller scale. Ferrous iron released by weathering may migrate laterally into valley floors, where in due course it becomes hardened. Subsequent erosion may expose such hardened units, which may finally stand higher than the surrounding country.

11.4.3 Etchplains and related features

This landform assemblage of plains and rocky outcrops can best be understood within the theory that landscapes may evolve as a response to chemical denudation, i.e. the removal of ions in solution. The idea has been applied particularly to tropical environments, where the consequences of weathering are perhaps most apparent. The theory involves the following features.

- The development of a weathered profile from which the more mobile elements have been removed.
- The formation of an etched rock surface below the weathering profile, and whose character is determined by the chemistry and structure of the parent material.
- A delicate relationship between the rate at which the regolith is removed, by physical (e.g. surface wash) as well as chemical processes, and the weathering rate at the bedrock surface ('basal weathering surface' or 'weathering front'). Fieldwork in Malawi has, however, shown little sign of surface wash on the 'African' laterised surface, and studies in Zimbabwe suggest that rainfall descends through rather than runs over

the same land surface. Weathering processes, on the other hand, involving the loss of both aluminium and silicon from kaolinite and leading to collapse of the saprolite, may be important in landscape reduction.

A land surface that results from large-scale chemical weathering and the possible development of etched rock surfaces is an *etchplain* (*see* Box 11.2 for an example). This term does not imply complete removal of the regolith. The land-forming process is called *dynamic etchplanation*. The nature of the

Box 11.2 A developing etchplain: the Koidu region of Sierra Leone

This is a granite-gneiss plateau at 370–410 m above sea level and bounded by steep-sided ridges and hills that rise to 810 m. The climate is hot, humid, and monsoonal, with an annual rainfall of some 2355 mm, 80 per cent of which falls in 6 months. The mean annual temperature is about 27°C. Until recently the region supported a semi-evergreen forest.

The landscape shows nearly-flat interfluves (50 per cent of the surface) and a drainage network of valley heads and valley swamps lacking channels (20 per cent). River valleys with channels make up only 20 per cent of the drainage system. The dominant clay mineral in the regolith is kaolinite.

Processes affecting the landscape include the following.

- *Bedrock etching* beneath interfluves and stream channels, with a local emergence of tors and corestones.
- *Transfer of ions in solution*, with some elements (iron and aluminium) being complexed by chelating substances. Reducing conditions in the swampy valley floors lead to the leaching of Fe^{2+} from nodules and fragments moved there by erosion.
- *Vertical transfer of clays* in interfluve regoliths, and *lateral transfer* from interfluves through valley heads to valley floors.
- *Periodic flushing-out* of coarse sediments along valley heads and valley floors during high discharge events, perhaps during periods of rapid climatic change during the Quaternary.

In summary, this is a landscape where chemical denudation is important. The active processes seem consistent with the idea of etchplanation, and are bringing about a progressive lowering, with the exception of locally resistant rocks and sites where duricrusts act as a protective cover.

(Based on material in Teeuw, R.M., Thomas, M.F. and Thorp, M.B. 1994: *Regolith and landscape development in the Koidu Basin of Sierra Leone*. In Robinson, D.A. and Williams, R.G. (eds) *Rock weathering and landform evolution*. Chichester: John Wiley, 303–20.)

balance between regolith development and removal allows a useful classification of etchplains:

- mantled etchplains, which may be active or fossil;
- partly stripped etchplains;
- stripped etchplains (etch surfaces);
- complex etchplains (weathered, incised or pedimented etch surfaces; in this context a pediment is an undissected slope at the foot of a mountain zone and cut across saprolite or other soft sediment);
- exhumed etch surfaces.

The key component of the etchplain theory is the weathering profile, and several types of profile behaviour may be recognised.

- *Profile lowering* involves a steady advance of the weathering front balanced (over 10^2–10^4 years) by the rate at which the land surface is lowered. If during weathering Fe^{3+} is precipitated as goethite which then evolves to haematite, a resistant horizon may form. This will tend to armour the regolith, especially when supported by a 'stone line' of fragments of resistant minerals such as vein quartz. Profile lowering is then replaced by profile deepening.
- *Profile deepening* takes place when weathering penetration continues but the land surface is largely protected from erosion. This is typically the case when a duricrust develops but free drainage is encouraged by uplift and dissection.
- *Profile thinning* occurs when the rate of surface lowering exceeds the rate of advance of the weathering front, perhaps because resistant rock is met at depth. The weathering profile becomes compressed. Finally a duricrust may rest on hardly altered rock.
- *Profile truncation* occurs when accelerated surface erosion, due perhaps to climatic change, removes the upper horizons. Less-evolved lower horizons are then exposed at the land surface.
- *Profile collapse* takes place when a kaolinite-rich regolith is exposed to freely drained conditions, and extensive leaching of Al^{3+}, Fe^{2+} and SiO_2 occurs. This leads to the breakdown of the kaolinite and the formation of a white, sandy residue.

Over a long time scale the behaviour of the profile is affected mainly by geological and hydrological factors. The *geological influences* include tectonic and structural components. When uplift and/or tilting occurs, the development of the profile may be accelerated. Uplift may lead to a thickening of the weathering profile as in the Sula Mountains, Sierra Leone, where the profile extends locally to about 130 m. Such uplift also affects the etchplain landscape as adjacent rocky massifs increase their relative relief compared to the adjacent down-wasting plains. The structural properties of rock guide the descending weathering front. Deep weathering penetration is favoured by the presence of jointed and fissile bedrock, and by the presence of shatter zones.

Hydraulic factors, essentially the position and movement of the water table, may be influenced over the longer term by geological controls such as uplift, but they make their own direct contribution to profile development and ultimately landscape character. Generally two circumstances are important. Conditions of impeded drainage, or of a fluctuating water-table, tend to be associated with the retention of mobilised cations and the concentration of oxides. A subsequent fall in the water table encourages the development of duricrusts and the deepening of weathering profiles below them.

Etchplain development, as measured by the rates at which land surfaces are lowered and weathering fronts descend, varies with the environment. Some calculated rates are

- 0.1 m Ma^{-1} (i.e. 0.1 metres per million years) for the interior region of the West Australian craton, and 1 m Ma^{-1} for its margin (a craton is a long-stable part of the Earth's crust, often made of old cystalline rocks);
- 6 m Ma^{-1} for the lowering of the Koidu etchplain, Sierra Leone;
- 20 m Ma^{-1} over a period of 4 Ma for the lowering of the weathering front in the Mojave Desert, USA;
- 25–38 m Ma^{-1} for the penetration of the weathering front into fresh rock of the humid, warm-temperate region of the southern Appalachians, USA;
- 50 m Ma^{-1} for the rate of denudation in equatorial northwest Kalimantan, Borneo.

The etchplanation mechanism, based on an acceptance of the importance of chemical denudation, provides a valuable approach to the understanding of tropical landscapes, and may well have wider application to other environments.

READINGS FOR FURTHER STUDY

You will find that most textbooks on landforms and surface processes have chapters on the contribution of weathering, e.g.

Selby, M. 1985: Weathering and landforms. In *Earth's changing surface*. Oxford: Oxford University Press, 189–209. A helpful introduction which also covers many other weathering topics.

More advanced material is to be found in

Robinson, D.A. and Williams, R.B.G. 1994: *Rock weathering and landform evolution*. Chichester: John Wiley. This volume contains much material, not only on landforms. Be selective in your approach.

Many of the articles referenced in this chapter will help you to explore ideas in more detail. Here are some additional readings on specific topics:

Fiol, L., Fornos, J.J. and Gines, A. 1996: Effects of biokarstic processes on the development of solutional rillenkarren in limestone rocks. *Earth Surface Processes and Landforms* **21**, 447–52. Puts forward the idea that the activity of a blue-green alga contributes to the formation of certain solution channels.

Gerrard, A.J. 1994: Classics in physical geography revisited: the problem of tors. *Progress in Physical Geography* **18**, 559–63. A review of a classical hypothesis of tor formation in the light of more recent investigations.

Swantesson, J.O.H. 1992: Recent microweathering phenomena in Southern and Central Sweden. *Permafrost and Periglacial Processes* **3**, 275–92. Contains a classification and discussion of the nature of microforms in crystalline rocks in a cool-temperate climate.

Thomas, M.F. 1989: The role of etch processes in landform development II. Etching and the formation of relief. *Zeitschrift für Geomorphologie* Suppl. N.F. **33** 257–74. A discussion of the theory of etchplanation, with special reference to tropical landscapes.

APPENDIX A NAMED COMPOUNDS AND MINERALS

Acetic acid (ethanoic acid)	CH_3COOH
Actimolite	$Ca_2(Mg, Fe)_5[Si_8O_{22}](OH, F)_2$
Albite	$NaAlSi_3O_8$
Alumina	Al_2O_3
Anatase	TiO_2
Andalusite	Al_2SiO_5
Anhydrite	$CaSO_4$
Anorthite	$CaAl_2Si_2O_8$
Apatite	$Ca_5(PO_4)_3F$ (this is for fluor-apatite, the common version)
Aragonite	$CaCO_3$
Augite	(a complex aluminosilicate so no formula given)
Bicarbonate ion	HCO_3^-
Birnessite	$(Na,Ca)Mn_7O_{14} \cdot 3H_2O$
Biotite (a trioctahedral mica)	$[(Mg,Fe)_3(Si_3Al)O_{10}(OH)_2]^-K^+$
Boehmite	γ-AlO(OH)
Bornite	Cu_5FeS_4
Brucite	$Mg(OH)_2$
Calcite	$CaCO_3$
Carbonic acid	H_2CO_3
Chalcedony	$SiO_2 \cdot nH_2O$
Chalcopyrite	$CuFeS_2$
Chlorite	$(Mg,Al,Fe)_{12}[(Si,Al)_8O_{20}](OH)_{16}$
Chrysocolla	$Cu_8Si_8O_{20}(OH)_8 \cdot 8H_2O$
Chrysotile	$Mg_3(Si_2O_5)(OH)_4$
Citric acid	$C(OH)(COOH)(CH_2COOH)_2$
Corundum	α-Al_2O_3
Cristobalite	SiO_2
Cryptomelane	KMn_8O_{16}
Cuprite	Cu_2O
Diaspore	α-AlO(OH)
Diopside	$Ca(Mg,Fe)[SiO_5]$
Dioptase	$CuSiO_3 \cdot H_2O$
Dolomite	$CaMg(CO_3)_2$

Enstatite	$Mg_2Si_2O_6$
Epidote	$Ca_2Fe^{3+}Al_2O(OH)(Si_2O_7)(SiO_4)$
Epsomite	$MgSO_4 \cdot 7H_2O$
Ethanoic acid	(*See* acetic acid)
Fayalite	Fe_2SiO_4
Flint	$SiO_2 \cdot nH_2O$ (1% H_2O present)
Fluorite	CaF_2
Formic acid (methanoic acid)	HCOOH
Forsterite	Mg_2SiO_4
Fulvic acid	(No exact formula – complex mixture of organic acids)
Gibbsite	γ-$Al(OH)_3$
Glushkinsite	(*See* Magnesium oxalate dihydrate)
Goethite	FeO(OH)
Gypsum	$CaSO_4 \cdot 2H_2O$
Halite	NaCl
Hematite	Fe_2O_3
Hornblende	$(Na,K)_{0-1}Ca_2(Mg,Fe^{2+},Fe^{3+},Al)_5$ $[Si_{6-7}Al_{2-1}O_{22}](OH,F)_2$
Humic acid	(No exact formula – complex mixture of organic acids)
Hypersthene	$(Mg,Fe^{2+})(SiO_3)$
Illite	$K_{1-1.5}Al_4(Si_{7-6.5}Al_{1-1.5}O_{20})(OH)_4$
Ilmenite	$FeTiO_3$
Iron pyrites	FeS_2
Jadeite	$NaAlSi_2O_6$
Jasper	A form of chalcedony
Kaolinite	$Al_2Si_2O_5(OH)_4$
Kieserite	$MgSO_4 \cdot H_2O$
Kyanite	Al_2SiO_5
Lactic acid	$CH_3CH(OH)COOH$
Limestone	$CaCO_3$ (*see* also dolomite)
Limonite	$2Fe_2O_3 \cdot 2H_2O$
Lithiophorite	$(Al,Li)MnO_2(OH)_2$
Magnesia	MgO
Magnesite	$MgCO_3$
Magnesium oxalate dihydrate	$MgC_2O_4 \cdot 2H_2O$
Magnetite	Fe_3O_4
Malachite	$Cu_2CO_3(OH)_2$
Malic acid	$C(OH)(COOH)(CH_2COOH)_2$
Manganite	$MnO \cdot OH$
Methanoic acid	(*See* formic acid)
Mirabilite	$Na_2SO_4 \cdot 10H_2O$
Monazite	$(Ce,La,Th)PO_4$
Muscovite (a dioctahedral mica)	$[Al_2(Si_3Al)O_{10}(OH)_2]^-K^+$
Natron	$Na_2CO_3 \cdot 10H_2O$
Olivine	$MgFeSiO_4$
Opal	$SiO_2 \cdot nH_2O$
Orthoclase	$KAlSi_3O_8$
Oxalic acid	$(COOH)_2$

Pyrites	(*See* iron pyrites)
Pyrolusite	MnO_2
Quartz	SiO_2
Rhodocrosite	$MnCO_3$
Rhodonite	$MnSiO_3$
Rock salt	$NaCl$
Rutile	TiO_2
Sericite	(*See* muscovite)
Serpentine	$H_4Mg_3Si_2O_9$
Siderite	$FeCO_3$
Silica	SiO_2
Silicic acid	H_4SiO_4 or $Si(OH)_4$
Sillimanite	Al_2SiO_5
Sodium feldspar	(*See* albite)
Sphalerite	ZnS
Sphene	$CuTi(SiO_4)(O(OH)F)$
Spinel	$MgAl_2O_4$
Spodumene	$LiAlSi_2O_6$
Staurolite	$(Fe^{2+},Mg)_2(Al,Fe^{3+})_9O_6[SiO_4]_4(O,OH)_2$
Thenardite	Na_2SO_4
Topaz	$Al_2(SiO_4)(OH,F)_2$
Tourmaline	(a complex aluminosilicate, so no formula given)
Tremolite	$Ca_2Mg_5Si_8O_{22}(OH)_2$
Vermiculite	(a complex smectite-like aluminosilicate)
Zircon	$ZrSiO_4$
Zoisite	$Ca_2Al(Al_2O)(OH)(Si_2O_7)(SiO_4)$

APPENDIX B
GLOSSARY

Acid: A proton (H^+) donor.
Actinomycetes: Soil-dwelling bacteria with cells arranged in rods or filaments.
Activation energy: The energy barrier to a chemical reaction proceeding.
Activity: Used in place of concentration to allow for the non-ideal behaviour of ions in solutions which are not dilute.
Aerobic: A description of an environment in which oxygen is present.
Aerosol: An arrangement of finely divided material suspended in a gas, for example very small salt particles in air.
Albedo: The extent to which a surface reflects solar radiation. It is expressed as a percentage of that radiation.
Alcrete: An alumina-rich horizon in a laterite. It may be hardened.
Alga (plural algae): Small and simple oxygen-producing green plants which have no roots, stem or leaves.
Alkali metals: The elements of group I of the periodic table which are lithium (Li), sodium (Na), potassium (K), rubidium (Rb) and cesium (Cs).
Allophane: A non-crystalline (amorphous) aluminosilicate containing varying amounts of Al_2O_3, SiO_2 and H_2O.
Aluminosilicates: Silicates in which some of the SiO_4 units have had the Si^{4+} replaced by Al^{3+}.
Amorphous: Used to describe a non-crystalline ('without shape') compound, e.g. amorphous silica (*see* allophane).
Anaerobic: A description of an environment in which oxygen is absent.
Andesite: A fine-grained volcanic igneous rock whose main component is plagioclase feldspar (up to 60–70 per cent).
Anhydrous: A system without water. Often used to describe salts, e.g. $CaSO_4$ is the anhydrous form of $CaSO_4 \cdot 2H_2O$.
Anion: A negatively charged ion.
Anisotropic: A substance which shows different characteristics depending on direction.
Aragonite: A crystalline form of calcium carbonate ($CaCO_3$).
Arène: A French term for a sandy regolith typically found on crystalline rocks and perhaps produced by moderate chemical weathering.
Atom: The smallest particle of an element. Consists of nucleus and surrounding electrons.
Atomic number: The number of protons found in the nucleus of an atom. The number also gives the position of the element in the periodic table of elements.
Basal weathering surface: The surface of unweathered parent material. It may be exposed by erosion to give an etchplain.

Base: A proton (H^+) acceptor. Often contains OH^-.
Bauxite: A residual product of intense chemical weathering, consisting largely of hydrated aluminium oxide with small amounts of iron oxide.
Bedding plane: A surface that separates beds in sedimentary rocks.
Biosphere: The zone which contains living organisms.
Biotic: Characterised by living organisms, including bacteria, fungi, plants and animals.
Bond polarity: An unequal sharing of electrons between atoms which are covalently bonded.
Calcite: A crystalline form of calcium carbonate ($CaCO_3$).
Calcrete: A type of chemical sediment that may be locally hardened. It consists chiefly of calcium carbonate in the form of calcite.
Capillarity: The tendency of a liquid to rise up a narrow rube.
Carbonation: The weathering effect of carbon dioxide in aqueous solution.
Cation: A positively charged ion.
Cation exchange capacity: The concentration of exchangeable cations expressed in millequivalents per 100 g of soil.
Cementing mineral: Mineral which binds together material, e.g. calcium carbonate, iron oxide minerals and silica.
Chelation: The complexing of an element by a ligand which has more than one point of attachment to the element.
Clay: A collection of particles smaller than 0.005 mm (5 micrometres) in diameter and which often includes clay minerals.
Clay mineral: A member of a class of very small, platy silicate minerals that are the product of chemical weathering.
Cleavage: Small-scale planes of weakness in a rock or mineral which are closely spaced and parallel to each other. Also the process of splitting along cleavage planes.
Climatic cabinet: An experimental device that allows rock samples to be exposed to controlled conditions of, for example, temperature and humidity.
Clitter: A deposit of coarse, angular debris produced by frost action and found round the bases of many Dartmoor tors.
Coefficient of thermal expansion: A measure of the amount of expansion a substance undergoes when heated. It is expressed as the fractional change in dimension per unit temperature change.
Complex: A chemical species in which coordination number is greater than oxidation number.
Compound: Two or more elements chemically combined together.
Compression wave: A wave that passes through a substance such as a rock when it is subject to a virtually instantaneous compression, e.g. by a hammer blow.
Conchoidal fracture: A curved 'shell-like' fracture surface that results when a hard homogenous material such as flint (SiO_2) is broken by a blow.
Corestone: A relatively small residual unit of rock, found above or below the land surface and rounded by chemical weathering.
Cortex: A skin of goethite (FeO(OH)) due to hydration, surrounding a hematite (Fe_2O_3) rich nodule in a laterite.
Covalent bonding: Chemical bonds in which pairs of electrons are shared between atoms.
Creep: Non-recoverable strain.
Cryohydrate: A mixture of water ice and salt crystals.

Crystalline texture: The texture shown by rocks which have interlocking crystals.
Crystallisation pressure: The pressure exerted on its surroundings by a crystal as it grows from solution.
Crystallographic axis: One of three axes (*a*, *b* and *c*) generally at right angles to each other and which are imaginary reference lines running parallel to the edges of the unit cell of a crystal.
Denudation: The removal of material from a landscape, as ions in solution (i.e. chemical denudation) and/or as mechanically transferred debris.
Dielectric constant: A measure of the ability of a solvent to reduce the forces between ions and thus to aid solution.
Dipole moment: A property possessed by a molecule in which there is an asymmetric distribution of electrons.
Dissolution: The process whereby ions pass from a mineral into a solution.
Duricrust: A hardened upper horizon of a tropical weathering profile. The main component may be an oxide of iron, aluminium or silicon.
Edaphogenic: Of the various factors, physical, chemical and biological, which contribute to soil formation.
Eh: A measure of the oxidising or reducing ability of natural waters.
Elastic limit: The maximum stress that a substance can withstand and still show elastic behaviour. Breakage may occur if it is exceeded.
Electromotive force: The relative tendency of electrons to be transferred to or from a chemical species expressed as a reduction potential (in volts).
Electron: A negatively charged particle found in an electron cloud surrounding the nucleus of an atom.
Element: Matter which contains all the same type of atoms.
Enthalpy: Transferred heat.
Enthalpy of solution: The heat change which occurs when 1 mole of a substance dissolves in a solvent.
Entropy: The randomness (or disordered state) of a system.
Equigranular: Descriptive of a rock whose component grains are of a similar size.
Equilibrium (chemical): A chemical reaction at equilibrium has constant amounts of reagents and products present.
Equilibrium (in weathering): A stable state shown by the persistence of resistant minerals. A partial or quasi-equilibrium is more usual.
Equivalent weight: The amount of a substance which will react with 1 equivalent weight (EW) of another substance. EW = RMM (relative molecular mass) for monobasic acids (e.g. HCl) and EW = RMM/2 for dibasic acids (e.g. H_2SO_4).
Etchplain: A plain produced in tropical regions when a chemically weathered regolith is removed by denudation.
Eutectic concentration: *See* eutectic temperature.
Eutectic temperature: The temperature at which a solution crystallises to give solid (e.g. salt) with the same concentration as the solution. This is called the eutectic concentration.
Fabric: The arrangement of minerals and particles in a rock.
Faults: Fracture surfaces in a rock along which displacement has occurred.
Feldspars: A group of aluminosilicates, mainly of calcium, sodium and potassium.
Ferrallitic: Descriptive of a weathering trend leading to an enrichment of iron and aluminium.
Ferricrete: The hardened upper zone of a laterite consisting largely of iron oxides (hematite and goethite).

Ferromagnesian: Descriptive of an earth material, especially a mineral, having significant amounts of iron and magnesium.
Fick's law: An expression which allows the rate of diffusion of gases to be calculated.
Fissile: Descriptive of a rock or mineral which is easily split along closely spaced weakness planes.
Flushing: The action of a flowing solution in carrying away ions from a weathering mineral.
Free energy: Energy that is available to do work.
Friable: Easily crumbled, particularly by hand.
Frost bursting: A fracturing of rock that occurs when supercooled water is suddenly converted to ice.
Frozen fringe: A transitional zone, of varying permeability, between unfrozen and frozen rock in a cold environment.
Fungus (plural fungi): A plant which lacks chlorophyll and obtains its nutrients from other plants or animals.
Gabbro: A coarse-grained igneous rock formed at depth. It is low in silica and high in magnesium and calcium.
Geomorphology: The study of the nature and origin of landforms, and of the processes acting on them.
Gibbs free energy: Named after J.W. Gibbs (*see* free energy).
Gneiss: A coarse-grained banded rock formed under conditions of high temperature and pressure.
Growan: A chemically weathered regolith found on Dartmoor.
Grus: Platy fragments of weathered granite, found typically in Dartmoor regoliths.
Half-life: The time taken for the concentration of a reagent to fall to half of its original value.
Halogens: The elements of group VII of the periodic table which are fluorine (F_2), chlorine (Cl_2), bromine (Br_2), and iodine (I_2).
Head: A regolith (q.v.) produced by frost action and thus dominated by angular debris. It may show evidence of downslope movement.
Heavy mineral: A mineral having a density greater than about $2.9\,g\,cm^{-3}$, e.g. hornblende, tourmaline and zircon.
Henry's law: The relationship of the solubility of a gas in a solvent and the partial pressure of the gas above the solvent.
Heterotrophic bacteria: Those which cannot use carbon dioxide as their only source of carbon and require one or more organic compounds.
Horizon: A layer within a soil or regolith that can be distinguished from adjacent layers by physical and/or chemical properties.
Hydration: The physical interaction of water with individual ions or with a solid, e.g. clay.
Hydration energy (enthalpy of hydration): The energy liberated when 1 mole of gaseous ions is passed into water.
Hydrodynamics: The study of the nature and rate of water movement, especially through rock and regolith.
Hydrogen bonding: A particular type of bonding which only occurs when hydrogen is bonded to an electronegative element. An important example is water.
Hydrolysis: Reaction with water in which an O–H bond is broken.
Hydronium ion (H_3O^+): Formed by the combination of H^+ and H_2O. It is a hydrated proton.
Hydroxide ion (OH^-): A negative ion which is a base.

Igneous: Descriptive of the rock produced when hot molten material from the Earth's interior cools either at depth or at the land surface.
Impermeable: Descriptive of a rock formation or regolith that barely allows water to pass through. Completely impermeable formations are rarely found.
Insolation: Exposure to the sun.
Interfluve: An area of land between two adjacent streams.
Interlayer (clay minerals): Loosely bound cations and/or layers of water molecules which occupy the space, of variable width, between layers of 2:1 clay minerals especially.
Inverted relief: A type of landscape where the low ground coincides with a structural 'high', e.g. an upfold, and vice versa.
Ion: An atom which carries a positive charge(s) (e.g. Na^+; a cation), or a negative charge(s) (e.g. O^{2-}; an anion).
Ionic bonding: Occurs in compounds in which positive and negative ions are held together by mutual attraction in an ionic solid.
Ionic potential: Obtained by dividing the charge on an ion by its radius in angstroms. It is an indicator of an ion's hydration energy and contribution to lattice energy.
Ionic strength: A measure of the total number of ions present per unit volume of a solution.
Isotope: Compared to another atom, it is an atom with the same number of protons, but with a different number of neutrons in its nucleus, e.g. ^{14}C and ^{12}C.
Isotopic dating: A method of determining the age of a specimen by finding out the extent to which an isotope in the specimen has decayed, e.g. ^{40}K and ^{14}C.
Isovolumetric method: A way of measuring the intensity of weathering, based on the assumption that there has been no loss of volume between an unweathered and weathered rock unit.
Joints: Small-scale fractures in a rock showing no displacement.
Kinetics: The quantitative study of the rates of chemical reactions.
Laterite: A regolith produced by intense weathering, and consisting mainly of goethite, hematite, aluminium hydroxides, kaolinite minerals and quartz, in varying proportions. Iron- and aluminium-rich laterites are found.
Lattice: A geometric pattern of atoms or ions or molecules in a crystal.
Lattice enthalpy: The amount of energy which is liberated when 1 mole of ionic solid is formed from gaseous ions coming together from infinity.
Layer (clay minerals): A combination of sheets, as when tetrahedral and octahedral sheets are bonded.
Leaching: The process by which ions are selectively removed in solution from a weathering substance.
Le Chatelier's principle: When a stress is applied to a system in dynamic equilibrium, the equilibrium adjusts to minimise the effect of the stress.
Lichen: A composite organism of a fungus and an alga living symbiotically.
Liesegang rings: Alternating brown and pale-coloured bands found in iron-rich rocks undergoing oxidation and reduction.
Ligand: A species which is attached to the central atom in a complex. It may be an ion, e.g. Cl^-, or a molecule, e.g. NH_3.
Lithodependence: Referring to a material, typically a regolith, having properties determined by the character of the parent rock.
Loam: A soil which has roughly equal proportions of sand, silt and clay.
Massif: An uplifted geological unit, typically made of hard rock and so standing above surrounding country, e.g. the Massif Central of France.

Mass number: The number of protons and neutrons in the nucleus of an atom.
Megacryst: A large crystal. Megacrysts of feldspar occur in the Dartmoor granite.
Metallogenium bacteria: Those which may oxidise metals.
Metamorphic: Descriptive of an earth material altered by intense heat and/or pressure.
Microcracks: Small-scale cracks (physical discontinuities) which may occur between and within mineral grains in rocks.
Microporosity: The proportion of micropores (very small pores) expressed as a percentage of total pore space.
Microtopography: Small-scale surface relief (variation in height across a surface).
Mineral species: A named mineral of distinctive character which is a member of a class. For example, the species quartz (SiO_2) is a member of the class of silicates.
Mixotrophic bacteria: Those which may utilise organic and/or inorganic nutrient sources.
Modulus of elasticity: The relationship between stress and strain for a material behaving elastically.
Molar mass: The mass of 1 mole of an element or a compound.
Molarity: Concentration of a solution measured in moles of solute per cubic decimetre of solution.
Molasse: A term used to describe undeformed sediments produced by the erosion of a mountain range when folding and uplift has been completed.
Mole (mol): The SI unit of amount of substance and equal to the number of atoms of ^{12}C in 12.0000 g of ^{12}C.
Molecule: Atoms bonded together by covalent bonding, e.g. I_2 and H_2O.
Monsoon: A seasonal wind, associated with contrasting weather conditions, in tropical and subtropical areas. In India, the southwest monsoon is associated with the rainy season.
Mottled zone: A zone or horizon in a laterite that consists of reddish-brown, non-rich mottles in a white or grey matrix that is dominated by quartz and kaolinite.
Neoformation: A deposit or substance which is a by-product of weathering processes, especially chemical.
Neutron: A neutral particle found in the nucleus of an atom.
Nitrobacteria: Those which convert ammonium ions to nitrates and nitrites.
Nucleus: The dense central part of an atom which contains protons and neutrons.
Octahedral: Having the shape of a regular octahedron which has six corners and eight faces, each of which is an equilateral triangle.
Ordered water: An arrangement of a sheet of water molecules whereby each molecule, aligned parallel to its neighbours, has its same end bound to an adjacent surface.
Organic acids: These contain the carboxylic acid group (–COOH), and in more complicated acids the phenolic group is present.
Oxidation: The loss of electrons by an atom or an ion. It involves an increase in oxidation number.
Pallid zone: A leached, pale-coloured weathering horizon which occurs below the water table in some tropical regoliths.
Peneplain: A land surface of low relief, perhaps showing a thick regolith (q.v.), and often developed across a variety of rocks and structures.
Periodic table: A classification of all of the elements into groups and periods.
Permeability: The ability of a rock to transmit a fluid.
Permeable: Descriptive of a rock formation or regolith that readily allows water to pass through.

Petrographic: Descriptive of rocks from hand-held samples to the microscopic scale.
pH: A scale of acidity and basicity. $pH = -\log_{10}[H_3O^+]$.
Phase change: A change in the state of a substance, as when liquid water changes to ice.
Phreatic zone: Area below the water table and therefore saturated with water.
Planation surface: A land surface of low relief produced by an unspecified erosional process.
Plane (clay minerals): A two-dimensional array of atoms, such as the basal oxygens of linked tetrahedra in clay minerals.
Podzol: A soil showing a grey, strongly leached, upper (E) horizon. Often found on well-drained, sandy rocks in humid temperate regions.
Poikilitic texture: A form of crystalline structure when a large crystal encloses one or more smaller crystals.
Porosity: The amount of void space in a rock measured as the volume of voids divided by total volume.
Pressure melting constant: A factor that describes the rate at which the freezing point falls with an increase in pressure. For ice it is 0.074°C MPa^{-1}.
Proton: A positive charged particle found in the nucleus of an atom.
Pseudo-bedding plane: A separation in an igneous rock that appears similar to a bedding plane in a sedimentary rock, but which is probably due to expansion.
Reactive site: A very small site at a mineral surface, e.g. a microfracture, where weathering is relatively rapid.
Reduction: The gain of electrons by an atom or an ion. It involves a decrease in oxidation number.
Reduction potential: *See* electromotive force.
Regolith: The mantle of unconsolidated material, showing varying degrees of weathering, which rests on solid, unaltered rock.
Regression equation: A mathematical description of the relationship (typically a straight line or curve) between two variables.
Relative atomic mass (RAM): The average mass of an atom of an element relative to one-twelfth of the mass of ^{12}C.
Relative molecular mass (RMM): The mass of 1 molecule of a compound relative to one-twelfth of the mass of ^{12}C. Obtained by adding together the RAMs of the atoms in the compound.
Rhizosphere: The upper part of the regolith dominated by roots.
Rhyolite: A fine-grained, acidic, igneous rock which often weathers to a creamy-white colour.
Rock quality designation (RQD): A value used in civil engineering to describe the spacing of the discontinuities (cracks) in a rock.
Saprolite: Chemically weathered material that lies beneath the upper zone of biological activity.
Saturation: The state of a solution when it can just maintain a solute without precipitation.
Saturation coefficient: The amount of water absorbed in 24 h when a sample is completely immersed, expressed as a proportion of total available pore space.
Savanna: An area of tropical or subtropical grassland, with few trees.
Schist: A metamorphic rock whose component minerals have a parallel arrangement.
Schmidt hammer: A device for determining rock strength, and thus intensity of weathering, by measuring rebound following a hammer blow.

Schorl: A black tourmaline produced when hot gases alter a granite shortly after its formation, as occurred in southwest England.
Sedimentary texture: Shown by sedimentary rocks and often lacks interlocking character (*see* crystalline texture).
Self-ionisation: The breaking down of a molecule, e.g. water, to give ions (hydronium ions and hydroxide ions for water).
Shatter zone: A zone or band of rock that has been fragmented by earth movements, often faulting.
Sheet (clay minerals): A three-dimensional array of linked atoms, such as octahedral or tetrahedral sheets in clay minerals.
Sheet joints: Joints trending parallel to a land surface and which may result from tensile stress due to removal of overlying material by erosion. Also called 'dilation joints'.
Silcrete: A hardened weathering formation showing a significant concentration of silica (silicon dioxide, SiO_2).
Silicates: Compounds which contain tetrahedral SiO_4 units.
Silt: A collection of particles ranging in diameter between 0.05 and 0.005 mm (5–50 micrometres).
Solubility product: A constant for the solubility of sparingly soluble salts and equal to the product of the concentration of the ions.
Solute: A substance which dissolves in a solvent.
Solvent: A liquid in which substances may dissolve.
Specific surface area: The surface area of, for example, a mineral, expressed in $m^2 g^{-1}$.
Spinels: A group of oxide minerals which contain 2+ and 3+ charged cations.
Spontaneous reaction: A reaction which occurs with decreasing free energy.
Stadial: A minor cold phase during an ice age when glaciers readvance. The Loch Lomond stadial affected Britain between 10 000 and 11 000 years ago.
Steppe: A (Russian) term for an extensive treeless plain.
Strain: Deformation resulting from stress.
Stress: Force per unit area.
Stress corrosion limit: A critical value of stress below which crack growth in a rock ceases.
Stress intensity factor: A measure of the stress that may be generated by suction along a free energy gradient in a water–ice system.
Supercooled water: Unfrozen water in a rock or regolith that is well below freezing point, perhaps by 5 or 6°C.
Supersaturation: The state of a solution when it holds an excess of solute without precipitation taking place.
Symbiotic: Two or more living organisms associated together for mutual benefit, e.g. lichen.
Taiga: (Russian) term for the cool-temperate circumpolar forest dominated by conifers.
Tectonic: Of earth movements which typically lead to the faulting, folding, and/or vertical movement of rock.
Ternary phase diagram: A two-dimensional figure drawn to show the composition of a three-component system where the components add up to 100 per cent.
Tertiary: A geological period that began about 65 million years ago and ended about 2 million years ago.
Tetrahedral: Having the shape of a tetrahedron, which is a figure with four corners and four faces, each of which is an equilateral triangle.
Texture: Relationship of mineral grains in a rock.

Thermal conductivity: A measure of the rate at which a substance conducts heat, measured in $W\,m^{-1}\,K^{-1}$.
Thermal shock: A sudden and large change in temperature, caused, for example, by a bush fire.
Transition point: The temperature at which one form of a substance (e.g. a less hydrated salt) is changed into another form (e.g. a more hydrated salt).
Unit cell: The smallest representative unit of an ionic solid.
Unit structure (clay minerals): The arrangement of planes, sheets and layers to form a distinctive clay mineral. For example, the unit structure of kaolinite consists of alternating tetrahedral and octahedral sheets.
Vadose zone: The zone above the water table, which is not permanently saturated.
van der Waals forces: Attractive forces between molecules.
Water absorption capacity: The amount of water that can be absorbed by a unit of rock in unit time.
Water table: The upper surface of the permanently saturated, or groundwater, zone.
Weathering front: The junction between unaltered parent rock and the overlying regolith.
Weathering profile: A vertical section through the regolith, and which may pass through two or more horizons.
Weathering ratio: The relationship between more stable and less stable oxides, and used as a measure of weathering intensity.
Weathering zone: *See* horizon.

APPENDIX C
UNITS, CONSTANTS AND CONVERSION FACTORS

This is a selective table containing information used in the book.

Basic SI units

metre (m)	(length)
kilogram (kg)	(mass)
second (s)	(time)
kelvin (K)	(temperature)
mole (mol)	(amount of substance)
ampere (A)	(current)

Some derived SI units

energy	joule	$J = kg\,m^2\,s^{-2}$
force	newton	$N = J\,m^{-1}$
power	watt	$W = J\,s^{-1}$
potential difference	volt	$V = J\,A^{-1}\,s^{-1}$
electric charge	coulomb	$C = A\,s$
thermal conductivity		$\lambda = J\,s^{-1}\,m^{-1}\,K^{-1} = W\,m^{-1}\,K^{-1}$

Some useful prefixes

mega (M)	10^6
kilo (k)	10^3
centi (c)	10^{-2}
milli (m)	10^{-3}
micro (μ)	10^{-6}
nano (n)	10^{-9}
pico (p)	10^{-12}

Some units

length	1 μm (micrometre) = 10^{-6} m
	1 nm (nanometre) = 10^{-9} m
	1 Å (angstrom) = 10^{-8} cm = 10^{-10} m
	1 pm (picometre) = 10^{-12} m
volume	1 L (litre) = 10^{-3} m^3 = 1 dm^3
mass	1 t (tonne) = 10^3 kg = 1 Mg

pressure	1 Pa (pascal) = $1\,\mathrm{N\,m^{-2}}$
	1 bar = $10^5\,\mathrm{N\,m^{-2}}$
	1 atm (atmosphere) = $101\,325\,\mathrm{N\,m^{-2}}$
energy	1 cal (calorie) = 4.184 J
dipole moment	1 D (debye) = 3.336×10^{-30} C m

Physical constants

mass of hydrogen atom	$m_H = 1.6734 \times 10^{-27}$ kg
mass of electrons	$m_e = 9.1091 \times 10^{-31}$ kg
Boltzmann constant	$k = 1.3805 \times 10^{-23}\,\mathrm{J\,K^{-1}}$
Planck constant	$h = 6.6256 \times 10^{-34}$ J s
Avogadro constant	$N_A = 6.0225 \times 10^{23}\,\mathrm{mol^{-1}}$
gas constant	$R = N_A k = 8.314\,\mathrm{J\,K^{-1}\,mol^{-1}}$
freezing point of water (ice point)	= 273.15 K = 0°C
pi	$\pi = 3.1416$
dielectric constant (for pure water)	= 82

APPENDIX D
PERIODIC TABLE OF THE ELEMENTS

	1	2	3	4	5	6	7	8	9	10	11	12	13	14	15	16	17	18	
	I	II											III	IV	V	VI	VII	O	
1s	1 H																	2 He	
2s	3 Li	4 Be											5 B	6 C	7 N	8 O	9 F	10 Ne	2p
3s	11 Na	12 Mg											13 Al	14 Si	15 P	16 S	17 Cl	18 Ar	3p
4s	19 K	20 Ca	21 Sc	22 Ti	23 V	24 Cr	25 Mn	26 Fe	27 Co	28 Ni	29 Cu	30 Zn	31 Ga	32 Ge	33 As	34 Se	35 Br	36 Kr	4p
5s	37 Rb	38 Sr	39 Y	40 Zr	41 Nb	42 Mo	43 Tc	44 Ru	45 Rh	46 Pd	47 Ag	48 Cd	49 In	50 Sn	51 Sb	52 Te	53 I	54 Xe	5p
6s	55 Cs	56 Ba	57 *La	72 Hf	73 Ta	74 W	75 Re	76 Os	77 Ir	78 Pt	79 Au	80 Hg	81 Ti	82 Pb	83 Bi	84 Po	85 At	86 Rn	6p
7s	87 Fr	88 Ra	89 **Ac																

*Lanthanum Series

4f	58 Ce	59 Pr	60 Nd	61 Pm	62 Sm	63 Eu	64 Gd	65 Tb	66 Dy	67 Ho	68 Er	69 Tm	70 Yb	71 Lu

**Actinium Series

5f	90 Th	91 Pa	92 U	93 Np	94 Pu	95 Am	96 Cm	97 Bk	98 Cf	99 Es	100 Fm	101 Md	102 No	103 Lr

INDEX